STUDENT UNIT GUIDE

# AS Mathematics
# UNIT 6683

D1806267

Unit 6683: Statistics 1

Peter Naylor

Philip Allan Updates
Market Place
Deddington
Oxfordshire
OX15 0SE

**Orders**

Bookpoint Ltd, 130 Milton Park, Abingdon, Oxfordshire, OX14 4SB
tel: 01235 827720
fax: 01235 400454
e-mail: uk.orders@bookpoint.co.uk
Lines are open 9.00 a.m.–5.00 p.m., Monday to Saturday, with a 24-hour message
answering service. You can also order through the Philip Allan Updates website:
www.philipallan.co.uk

© Philip Allan Updates 2007

ISBN 978-1-84489-578-6

This guide has been written specifically to support students preparing for
the Edexcel AS Mathematics Unit 6683 examination. The content has been
neither approved nor endorsed by Edexcel and remains the sole responsibility
of the author.

Printed by MPG Books, Bodmin

Philip Allan Updates' policy is to use papers that are natural, renewable
and recyclable products and made from wood grown in sustainable forests.
The logging and manufacturing processes are expected to conform to the
environmental regulations of the country of origin.

AS Mathematics

# Contents

## Introduction

■ ■ ■

## Content Guidance

■ ■ ■

## Questions and Answers

# Introduction

## About this guide

This guide is for students following the Edexcel specification for AS and A2 mathematics. It deals with the first statistics module, **Unit 6683: Statistics 1**. It offers advice for the effective study of the module. The aim of the guide is to help you understand the key concepts and to see the ways in which some of the main ideas are interlinked. The guide has three sections.

- **Introduction** — this provides guidance on study and examination preparation, offering advice on examination skills. There are suggestions and points to note so that you can answer questions efficiently and effectively in the time that you are allocated for the examination.
- **Content Guidance** — this section does not attempt to cover all the theory behind the statistical concepts. It does, however, cover all the main features of the topics in the module, together with guidelines on how to ensure that your studies are fully rewarded.
- **Questions and Answers** — this shows you the sort of questions you are likely to be faced with in the examination. Solutions are provided to all the questions, together with a mark scheme to give some idea of how the marks are allocated. For some questions there are comments about the kinds of errors that are most often made and how to avoid them. For other questions there is some indication of the typical responses that a grade-C candidate might have made that pulls the solution down from an A or B grade.

A full understanding of this module requires proper attention to detail. It takes time to absorb information and to convert the basic knowledge into a full understanding. There are one or two statistical topics in the module that use some basic mathematics, such as solving simultaneous equations. Any initial problems with the material can be overcome if time and effort are invested.

It is worth noting that a good understanding of basic statistics can be of great assistance in other academic fields. Subjects such as biology, geography, economics, business studies and marketing often have some element of research and/or analysis in which statistics is an essential tool.

## The specification

The specification gives details of the topics likely to be tested in the examination, together with a list of formulae that need to be learned. Some formulae are printed in the booklet of formulae and statistical tables, which is issued to you at the beginning of the examination. The formulae that appear in the examination booklet are highlighted in the relevant sections of this guide. The full specification can be downloaded from the internet at www.edexcel.org.uk

# How to use this guide

This guide is designed to help you with your revision. You can revise a subject only if you have already studied it earlier in the course, so it is important that you are already familiar with the main ideas, having invested time and energy in learning the concepts before beginning your revision.

As you look through the content guidance, obviously you will pay attention to the main points in each section. However, you also need to look for hints on the little things that make the difference between a reasonable examination script and a good one.

I am sure you will find that some of the methods demonstrated in this guide are different from those you have been taught. After all, if you pick up two statistics textbooks, the methods for dealing with some of the concepts could be completely different. If you do come across a technique or method that is new to you, then have a look at it. However, if you prefer the method you are already familiar with, so long as it is likely to guarantee a successful solution for you, then the advice is to stick with it.

When you are using the Question and Answer section of the guide, it is suggested that you try to solve each question first, by covering up the solution. This enables you to check the main points of your answer with the given solution. The mark scheme should give you some idea of how many marks you would lose if your solution is not completely correct.

# Revision planning

We all have our own preferred methods for revision, but common to all successful outcomes is that revision takes time and commitment. When planning your revision, keep it clearly in mind that you are already likely to be better at some topics than others. This should focus your attention on how much time needs to be spent on each part of the specification. Your stronger topics will need some work in order to build on those strengths; your weaker areas may need much more time so that you understand the basic concepts and can develop your ability to transfer these basic ideas into the more demanding problems.

Remember that you cannot work all the time. You must build into your schedule some time for relaxation and any other commitments you have.

## Targets

When planning your revision programme you should set yourself short-term and long-term targets.

A short-term target may be to study a small section of a topic that you know you can cover in a particular revision period. For example, your target may be to spend half an hour trying to understand how the given class limits on a frequency table relate to the true class boundaries.

This, in turn, may be part of a long-term target to cover all aspects of graphical representation of data and calculations of averages and measures of spread. It is likely to be more productive if you set these targets for yourself, rather than simply sitting down with the aim of doing an hour's worth of statistics.

A useful short-term target is to make a list of the formulae that need to be learned, in other words the formulae that are not in the examination booklet of formulae and statistical tables.

## Textbooks

Textbooks are a valuable resource. They are likely to be helpful in your revision:
- for consolidating the important concepts
- for giving you access to questions that you can work through for extra practice
- for reminding you about topics from GCSE mathematics that might be needed in some areas of work

When using a textbook, it is important to focus on what you are trying to get out of the work. If you are refreshing yourself on how to draw a tree diagram in probability, ensure that you have a pen and paper to hand, for noting key points and practising important concepts as you read.

## Questions from past papers

Using real questions that have appeared on previous examination papers is likely to be one of the most valuable and important activities in your revision programme. You should try to obtain copies of both the papers and the mark schemes, so that you can check your solutions and identify where extra work is most needed. Looking at complete papers will also give you a good idea of the style of the questions and the typical balance of marks awarded on each topic.

With any of the modules for AS or A2 mathematics, the most effective way to revise is to answer questions and to check your solutions, correcting them and reviewing the topics where necessary. The more questions you tackle, the more marks you are likely to score in the examination. Although you will probably need to look through your own notes and the information and help in the content guidance section of this book, actually doing the calculations and interpreting the results is the most beneficial way to proceed.

# The modular examination

The examination lasts for 90 minutes and comprises about 7 questions. The marks for each question are given on the paper. The paper also contains the space for your solutions. When questions ask you to draw a graph, graph paper will be included. You are asked to attempt all the questions.

You must have a scientific calculator for this module. Calculators with a facility for symbolic algebra, differentiation and/or integration are not permitted. Some programmable calculators are permitted.

The examination is designed to allow you the opportunity to show what you know and understand. The mark scheme is constructed in such a way that the examiners are always looking for the opportunity to award the marks; they are not trying to find ways to penalise you.

The content of the specification is such that almost all topics appear on every paper. Certainly all the major topics will be included. It is worth noting that if you make the decision not to revise a particular topic because you find it too difficult, you are automatically forfeiting a certain number of marks. Do your best to overcome any problems with particular concepts, so that you can at least make a start on every question.

There are no prerequisites for this examination, but a good understanding of some topics from GCSE mathematics, such as stem and leaf diagrams and simultaneous equations, is needed.

The paper has five assessment objectives (AOs).

AO1 makes up 20–25% of the test. You should be able to:
- recall, select and use your knowledge of mathematical facts, concepts and techniques in a variety of contexts

AO2 makes up 20–25% of the test. You should be able to:
- construct rigorous mathematical arguments and proofs through use of precise statements, logical deduction and inference and by the manipulation of mathematical expressions, including the construction of extended arguments for handling substantial problems presented in unstructured form

AO3 makes up 15–20% of the test. You should be able to:
- recall, select and use your knowledge of standard mathematical models to represent situations in the real world
- recognise and understand given representations involving standard models
- present and interpret results from such models in terms of the original situation, including discussion of the assumptions made and refinement of such models

AO4 makes up 5–10% of the test. You should be able to:
- comprehend translations of common realistic contexts into mathematics

- use the results of calculations to make predictions, or comment on the context
- where appropriate, read critically and comprehend longer mathematical arguments or examples of applications

AO5 makes up 5–10% of the test. You should be able to:
- use contemporary calculator technology and other permitted resources (such as formulae booklets or statistical tables) accurately and efficiently
- understand when not to use such technology, and its limitations
- give answers to appropriate accuracy

## Key commands

Examiners use certain words that require you to answer in a specific way. These commands are often an important clue, giving you some idea of how much you need to write down.

### Calculate or evaluate
Some numerical work is needed.

### Write down
The required answer needs no detailed calculation. It ought to be quick and easy to work out.

### Show that
You need to show all the necessary working because the mark scheme will be designed to award marks for the method and not the given answer.

### Estimate
You must use your statistical knowledge to give the most realistic answer to a given calculation.

### Comment
Look at the answer you have already found and make a written statement about it.

### Compare and contrast
Look at the two sets of data or two graphs and write comments that relate to both sets.

### Give an interpretation of
Write a comment, in the context of the question, that describes what a particular result tells you.

### Give a justification of
Show some evidence that demonstrates a statement to be true. This may be some figures or the result of a calculation.

### Find the exact value of
Do not round off your answer but leave it, for example, as a fraction.

# Examination technique

It is also important to think about how you tackle the examination itself.

- *Read through the whole paper before you start.* The paper is designed so that the easier questions appear at the beginning. Nevertheless, you should look through all the questions first and then start with those that are on your stronger topics.
- *Read the question carefully.* Questions do vary from one paper to the next, so make sure that you understand the demands of the particular question you are about to answer.
- *Be aware of the number of marks for each question.* Pay particular care to questions that ask you to write something, rather than carry out a calculation. For example, if three marks are available, you need to write three points. If possible, try to write one or two extra points in case the ones you write first are not worth all the marks available.
- *Check the degree of accuracy required.* If a question asks for '2 decimal places' or '3 significant figures', you will lose a mark if you ignore the instruction. It is worth underlining such instructions before you start the question, to remind yourself of the need for appropriate accuracy. If the question does not ask for a fixed number of decimal places or significant figures, give your answer correct to 3 significant figures.
- *Show your working.* Some candidates do most of their working on a calculator and write down only the final answer. This is risky. If the final answer is wrong and you have written down nothing else, you automatically score zero, no matter how close you are to the correct answer. For example, consider a question that asks for an answer correct to 1 decimal place and the mark scheme gives an answer of 23.7. If you write 23.6 (possibly because you simply forget to round up the full answer of 23.681), then your answer will score zero, when it might have been worth 5 or 6 marks with appropriate and correct working shown.
- *Do not round off your values too early in any calculation.* If you round off and then use the rounded value, all the remaining answers will be inaccurate and hence marks will be lost.
- *Label graphs fully.* Axes on graphs need to be labelled correctly. Without the labels the graphs are meaningless.
- *Choose sensible scales when drawing graphs.* For example, never choose a scale that increases in multiples of 3. If your axis goes up 3, 6, 9, 12,..., you will lose the marks for a sensible choice of scale, because it is difficult to read a graph with this scale. You are also likely to lose one of the marks for plotting the graph, because it is difficult to check the accuracy when the scale is difficult to read.
- *Draw diagrams.* Some questions ask you to draw a diagram and certain topics lend themselves to using a diagram to help understand a particular context. For example, you should draw a diagram every time you attempt a question on the normal distribution. This will help you to avoid the most common mistakes.
- *Be prepared to write comments and not just carry out calculations.* As mentioned above, you might be asked to interpret a value that has been calculated. Your

interpretation needs to be in the context of the question and not simply a quotation from a statistics textbook. For example, if you are asked to interpret the correlation coefficients, don't just write, 'It shows a good positive correlation.' Explain what correlates with what; for example, 'People who score well in test 1 appear to score better in test 2.'

- *Guess.* You may be attempting a longer question in which you have to use an answer from a previous part of the question. If you are unable to do the earlier part, it is worth inventing an answer that you can use in the next part. For example, if you need to calculate a standard deviation in part (a) and you cannot do that calculation but need the standard deviation in part (b), simply write down a sensible value for the standard deviation and use it in part (b). You may still lose the accuracy marks in part (b), but you will almost certainly get the marks for a correct method. In some cases you may even get the accuracy marks if the mark scheme allows a lot of 'follow through' from your wrong answer to part (a).

## Some technical points

When you have completed your examination paper, it is sent to a processing centre where it is scanned so that examiners can mark the questions online. This raises some important points about how you should present your answers.

- You must write your answers using a *black* or *dark blue* pen. If you write in pencil or other coloured pen, there is a good chance that nothing will be visible on the screen once your solution has been scanned. If the examiner cannot see your answers, then obviously you will score zero.
- Ensure that all your solutions are written within the frames on the pages of your answer book. The margins do not appear on the scanned page and hence cannot be seen by the examiner.
- Forget all the rules you might have been taught about drawing graphs using a fine pointed pencil. If you use a thin, sharp pencil, it is highly likely that your graph will not be seen at all on-screen. Use a dark pen for graphs so that your answer is clearly visible.
- If you run out of space for a question, you have two options. Either insert an additional sheet of paper containing your completed solution or finish the question elsewhere in the answer book, but if you do this you must write a note to tell the examiner where the question continues. This is because the examiner will never see your entire paper. He/she will see only the pages relevant to a particular question. If the examiner sees a note from you stating that the question has been continued elsewhere, he/she will inform the senior examiners that they need to look at your solution as a special case.

Content
Guidance

This section is a guide to the content of **Unit 6683: Statistics 1**. It does not constitute a textbook for this unit.

The main areas of this unit are:

- **Mathematical models in probability and statistics** — what is a model?
- **Representation of sample data** — types of data, stem and leaf diagrams, grouped frequency distributions, histograms.
- **Methods for summarising sample data** — mode, mean, median, quartiles, box and whisker diagrams (box plots), range, interquartile range, variance and standard deviation, skewness.
- **Probability** — elementary probability, sample space, Venn diagrams, addition and multiplication rules, conditional probability, tree diagrams, independent and mutually exclusive events.
- **Correlation and regression** — scatter diagrams, product–moment correlation coefficient, least-squares regression line.
- **Discrete random variables** — probability functions, cumulative distribution functions, expectation and variance for a discrete random variable, the algebra of expectation and variance, the discrete uniform distribution.
- **The normal distribution** — properties of the normal distribution, standardisation, use of normal distribution tables, percentage points of the normal distribution.

For each part of the specification, you should also consult a standard textbook for more information.

Statistics is considered to be a branch of mathematics but it is a vital analytical tool. The examination requires you to perform many calculations, but you must remember that the purpose of any such calculations is to provide evidence from which conclusions can be drawn. Appropriate selection of statistical techniques is an essential quality for a good statistician. In the examination you will be asked to draw conclusions from some of your calculations, make comments and offer interpretations of your analysis.

# Mathematical models in probability and statistics

In this module only a simplistic understanding of modelling is required. The purpose of modelling is to help solve problems in the real world without the need to construct a physical model. Examples of situations that could use modelling for analysis are:

- Will making a road a dual carriageway reduce the number of accidents on a particular road?
- Is there a difference between the rates at which cars arrive at a filling station at the weekend compared with a weekday?
- Is there a difference in sentence length between two or more newspapers?

The stages in the modelling process are:

**(1)** Recognise a real-world problem.
**(2)** Devise a statistical model.
**(3)** Use the model to make predictions about what is likely to happen in the real-world situation.
**(4)** Collect experimental data.
**(5)** Obtain expected outcomes.
**(6)** Use statistical tests to see how well the model works, compared with the real world.
**(7)** Refine the model if necessary.

It is important to learn these seven stages.

Any model is a simplification of the real situation, so the advantages are that it is cheaper and quicker to produce than the real situation.

# Representation of sample data

## Types of statistical data

The different types of statistical data can be described as follows:

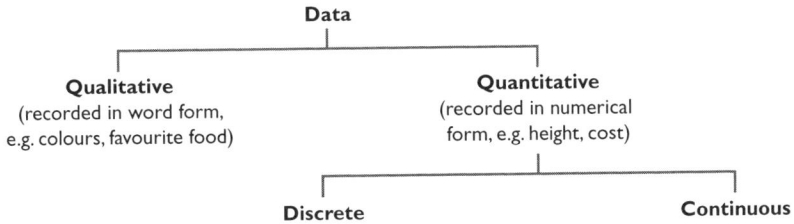

Little statistical work can be carried out on qualitative data. Unit 6683 is concerned with *quantitative* data.

**Discrete data** are usually recorded by **counting**. This kind of data can only take particular values. Examples are numbers of children in classes, shoe sizes, prices of chocolate bars or numbers of goals scored in football matches.

**Continuous data** are recorded by some form of **measurement**. This type of data can take any value within a certain range. Examples are heights of students, times taken to solve a problem, ages in years or weights.

# Stem and leaf diagrams

Stem and leaf diagrams can be used to represent data. You are unlikely to be asked to draw a stem and leaf diagram in the Unit 6683 examination, because this topic appears on the GCSE specification. You are more likely to be presented with a stem and leaf diagram and asked to read some information from it. When extracting information from the diagram, it is important to take note of the key and to remember that the value in the stem, as well as the value in the leaf, forms part of the answer.

The stem and leaf diagram below shows the number of television programmes that a group of children watched over a week.

```
0 | 7  8
1 | 0  2  2  3  5  8  9
2 | 1  4  5  8
3 | 2  6                    Key: 1 | 4 means 14 programmes
```

In this example the highest number of programmes watched is 36. It is a common mistake for candidates to forget to include the '3' from the stem and simply put the value '6'.

# Grouped frequency distributions

In later sections of this unit, data are likely to be presented in the form of a grouped frequency table. At this stage it is important to understand how the grouped data have been arranged into class intervals.

The following table shows the times taken, to the nearest minute, for the pupils in a class to solve a problem. The 5–9 interval needs to be looked at closely.

| Time taken/minutes | Frequency |
|---|---|
| 0–4 | 3 |
| 5–9 | 7 |
| 10–14 | 10 |
| 15–19 | 4 |

A lot of information is needed to appreciate the true meaning of the interval.

In any statistical analysis of these data, the **true boundaries** to each interval must be understood. In turn, this allows you to find the true **class width** and the **mid-point** of each interval.

The diagram below shows the key features of the 5–9 minute interval.

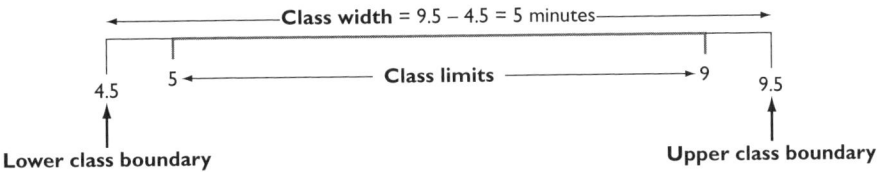

It is important to distinguish between the class *limits* and the class *boundaries*. In any graphical representation of the data, and in any calculations carried out using the data, it is the class boundaries that are essential to the accuracy of the analysis.

From the diagram, you can see:
class width = upper class boundary – lower class boundary

The mid-point of the interval is found using the **class boundaries**. In this case the mid-point is:
$\frac{1}{2}(4.5 + 9.5) = 7$

# Histograms

If data are **continuous**, they can be represented by a **histogram**. The key features of a histogram are:
- **Area is proportional to frequency.**
- The vertical axis represents **frequency density.**
- There must be **no gaps between the columns**, so each column must be plotted using the true class boundaries, not the class limits.

When drawing a histogram, the simplest method is to make the area of each column *equal* to the frequency.

To calculate the frequency densities (or heights of columns) use the following formula:

$$\text{frequency density} = \frac{\text{frequency}}{\text{class width}}$$

The grouped frequency table should be expanded to incorporate a column for the class widths and a column for the calculated frequency densities.

The data in the following table represent the lengths of worms collected in a garden. The columns that are needed in order to draw a histogram have been included.

| Length of worm/cm | Frequency | Class width | Frequency density |
|---|---|---|---|
| 1–5 | 8 | 5 | 8 ÷ 5 = 1.6 |
| 6–10 | 6 | 5 | 6 ÷ 5 = 1.2 |
| 11–15 | 5 | 5 | 5 ÷ 5 = 1 |
| 16–25 | 1 | 10 | 1 ÷ 10 = 0.1 |

In the example above, the class boundaries are 0.5–5.5, 5.5–10 etc.

When considering the true class boundaries, the highest value in one interval is *always* the same as the lowest value in the next interval. This ensures that there are no gaps between the columns.

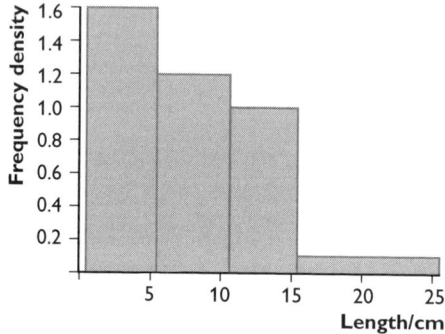

An alternative method for numbering the horizontal axis, and one that makes it easier to draw accurately, is shown below:

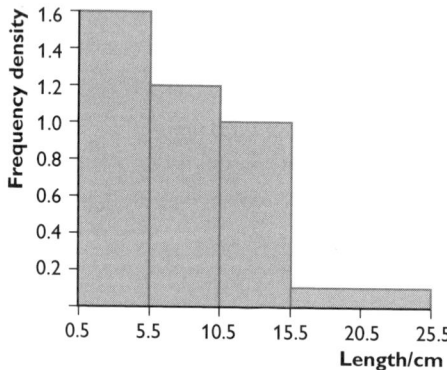

The frequency densities can be adjusted provided that they are still in the correct proportions. For example, multiplying all the above frequency densities by:
- a scale factor of 10 would give new frequency densities of 16, 12, 10 and 1
- a scale factor of 5 would give 8, 6, 5, 0.5

These alternative sets of values and many others would be equally acceptable on a correct histogram for the data above.

# Methods for summarising sample data

## Measures of location or average

### Mode

The **mode** is the value of the variable that occurs *most often*.
- In a **frequency distribution**, the mode is the value that has the **highest frequency**.
- In a **bar chart**, the mode is the value represented by the **highest column**.
- In a **grouped frequency table**, the **modal class** is the interval that has the **highest frequency**.

### Median

With the data arranged in ascending order, the median is the **middle value**.

The position of the median for $n$ values is always found by using the rule $\frac{n+1}{2}$.

- If there are 25 values, the median is $\frac{25+1}{2} = 13$th value.

- If there are 20 values, the median is $\frac{20+1}{2} = 10\frac{1}{2}$th value. This means that the median is halfway between the 10th and 11th values.

In this latter case, with an even number of values, *adding the middle pair* of values and *dividing by 2* gives the median.

A **cumulative frequency column** (found by adding the frequencies in turn as you move down the table) is needed if the data are presented in tabular form.

| $x$ | Frequency | Cumulative frequency |
|-----|-----------|----------------------|
| 4   | 12        | 12                   |
| 5   | 17        | 12 + 17 = 29         |
| 6   | 25        | 29 + 25 = 54         |
| 7   | 18        | 54 + 18 = 72         |
| 8   | 13        | 72 + 13 = 85         |
| 9   | 8         | 85 + 8 = 93          |
| 10  | 4         | 93 + 4 = 97          |

The median is $\frac{97+1}{2} = 49$th value.

To find the 49th value, look for the first number in the cumulative frequency column that is higher than 49. In this example it is 54, so the median is the value from the $x$ column that has 54 as its cumulative frequency. Here the median is 6.

If the data are presented in the form of a grouped frequency table, one method commonly used for estimating the median is to construct a cumulative frequency graph but this method is not examined at AS.

The method of **linear interpolation** is commonly tested. It assumes that the values within each class interval are equally spread across the interval and, as such, the calculation is based on proportions.

| x, weight/kg | Frequency | Cumulative frequency |
|---|---|---|
| 0–4 | 8 | 8 |
| 5–9 | 14 | 8 + 14 = 22 |
| 10–14 | 23 | 22 + 23 = 45 |
| 15–19 | 15 | 45 + 15 = 60 |
| 20–24 | 13 | 60 + 13 = 73 |
| 25–29 | 8 | 73 + 8 = 81 |
| 30–50 | 4 | 81 + 4 = 85 |

The median is $\dfrac{85 + 1}{2}$ = 43rd value.

Here the cumulative frequency, 43, and the median, $m$, are proportional to their respective intervals, so:

$$\frac{p}{14.5 - 9.5} = \frac{q}{45 - 22}$$

$$\frac{m - 9.5}{14.5 - 9.5} = \frac{43 - 22}{45 - 22}$$

Note that the **true class boundaries** are used in the calculation, not the class limits from the table.

$$\frac{m - 9.5}{5} = \frac{21}{23}$$

$$m - 9.5 = \frac{5 \times 21}{23}$$

$$m = 9.5 + \frac{5 \times 21}{23} = 14.1 \text{ kg}$$

## Quartiles

The **quartiles** for a set of data split the given values into four quarters. The notation used for the quartiles is usually $Q_1$ for the lower quartile and $Q_3$ for the upper quartile; the second quartile is the **median**.

There are several ways to locate $Q_1$ and $Q_3$. The simplest method for finding $Q_1$ is to consider all the values that lie *below* the median and find the middle value from this list. So for the 13 values below:

$$3 \quad 4 \quad 6 \quad 9 \quad 13 \quad 14 \quad 14 \quad 18 \quad 20 \quad 26 \quad 28 \quad 29 \quad 32$$

the median is the 7th number, found from $\dfrac{13+1}{2}$.

$$3 \quad 4 \quad 6 \quad 9 \quad 13 \quad 14 \quad \boxed{14} \quad 18 \quad 20 \quad 26 \quad 28 \quad 29 \quad 32$$

Median = 14

There are six values below the indicated median. To find the lower quartile, $Q_1$, we need the middle value from this list of six numbers. Here there is a middle pair, so $Q_1$ lies halfway between these two middle numbers: $Q_1 = 7.5$.

Similarly, the upper quartile, $Q_3$, is halfway between 26 and 28, so $Q_3 = 27$.

Once the quartiles have been found, the **interquartile range** (IQR) can be found using:

$\quad$ IQR = $Q_3 - Q_1$

In this example:

$\quad$ IQR = $Q_3 - Q_1 = 27 - 7.5 = 19.5$

## Box and whisker diagrams or box plots

The quartiles are commonly used to draw a **box and whisker diagram**.

The five key values required for drawing a box plot are the lowest and highest values from the data, plus the median and quartiles. It is essential to draw a linear scale below the box plot and to label the diagram fully. Examples can be seen in questions 3 and 4 in the Question and Answer section of this book (pp. 51–52).

## Outliers

If the data include any extreme values, it may be appropriate to calculate whether these data points are to be classed as **outliers**, before the box plot is drawn. If such calculations are required in an Edexcel examination question, the particular rule to be used will be stated in the question. The calculations to find any outliers must be shown clearly in your working.

The most common method for finding outliers is as follows:
- At the lower end of the distribution, an outlier is any value that is more than $1.5(Q_3 - Q_1)$ below $Q_1$.

- At the upper end, an outlier is any value that is more than $1.5(Q_3 - Q_1)$ above $Q_3$.

For example, if $Q_1 = 27$, $Q_3 = 41$, median $= 32$, lowest data point $= 9$ and highest two values are 60 and 84, then the calculations for the outliers are:

$$1.5(Q_3 - Q_1) = 1.5(41 - 27) = 1.5 \times 14 = 21$$

- Lower end: $Q_1 - 21 = 27 - 21 = 6$

There are no values below 6 and so there are no outliers at the lower end.

- Upper end: $Q_3 + 21 = 41 + 21 = 62$

84 is above the upper-end value, so 84 is an outlier.

It is now possible to draw the box plot.

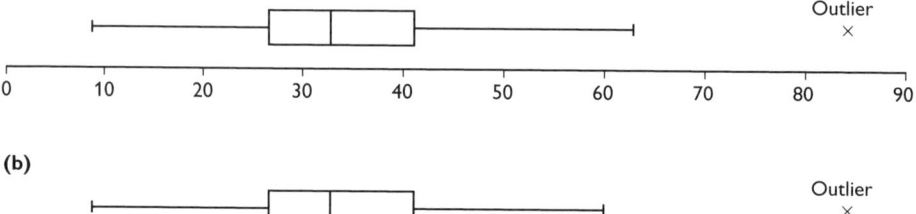

Note that the upper whisker does not go up to the outlier value. It stops at one of two possible values, both of which are equally acceptable. In plot (a) it stops at 62 (the critical value calculated at the upper end), whereas in plot (b) it stops at 60 (the highest data point that was not classified as an outlier).

## Purpose of a box plot

A box plot can be used to *comment on* and *interpret* the data from which it came. A box plot is particularly useful if you are trying to *compare* or *contrast* the data from two different sets, so you have two box plots drawn on the same grid, using the same scale.

When making such comparisons, you can see which of the two box plots has:
- the greater median (or state that the medians are equal)
- the larger interquartile range (or are they equal?)
- the larger range (or are they equal?)

You can also compare the **skewness** of the two sets of data. For example, if the median is nearer to $Q_1$ than to $Q_3$, the data set shows a positive skew.
- If $Q_2 - Q_1 < Q_3 - Q_2$, it has a positive skew.
- If $Q_2 - Q_1 > Q_3 - Q_2$, it has a negative skew.
- If $Q_2 - Q_1 = Q_3 - Q_2$, it is symmetric.

More information about skewness is given on p. 25.

It is worth noting that a question asking you to make comparisons between two box plots will state how many marks are available. If there are 4 marks, aim to make *four* relevant comparisons.

## Mean

For any set of data, the mean is found by adding the values and dividing by the number of values. For simple data, the mean, $\bar{x}$, is given by:

$$\bar{x} = \frac{\sum x}{n}$$

where $n$ is the number of values in the data set.

From a frequency distribution, this formula becomes:

$$\bar{x} = \frac{\sum fx}{\sum f}$$

This topic is covered in all GCSE specifications and detailed examples can be found in any GCSE textbook.

Here we look at how the ideas about true class boundaries and midpoints link in with calculating a mean from a grouped frequency distribution.

The following table represents the lengths of leaves (measured to the nearest centimetre) collected from a particular type of tree.

| Length/cm | Frequency, $f$ | True class boundaries | Mid-points of intervals, $x$ | $fx$ |
|---|---|---|---|---|
| 12–15 | 14 | 11.5–15.5 | 13.5 | 189 |
| 16–18 | 21 | 15.5–18.5 | 17 | 357 |
| 19–29 | 5 | 18.5–29.5 | 24 | 120 |
| | $\sum f = 40$ | | | $\sum fx = 666$ |

$$\text{mean, } \bar{x} = \frac{\sum fx}{\sum f} = \frac{666}{40} = 16.65 \text{ cm}$$

# Measures of dispersion or spread

## Range

The range is the simplest of the measures of spread:

range = highest value – lowest value

## Interquartile range (IQR)

IQR = upper quartile − lower quartile

or

$$IQR = Q_3 - Q_1$$

## Variance and standard deviation

Variance and standard deviation are the most important measures of spread. They are closely related because:

standard deviation = $\sqrt{\text{variance}}$

The principle behind the variance is that it considers the average distance from the mean of each of the original $x$ values. The steps in the process for finding the variance are:

- Step 1: Measure the deviations from the mean of each of the original $x$ values.
  $x - \bar{x}$
- Step 2: Square all these deviations.
  $(x - \bar{x})^2$
- Step 3: Find the sum all of these squared deviations.
  $\sum (x - \bar{x})^2$
- Step 4: Hence find the average of these squared deviations.
  $$\frac{\sum (x - \bar{x})^2}{n}$$

Hence the formula for the variance is:

$$\text{var} = \frac{\sum (x - \bar{x})^2}{n}$$

This is the theoretical formula. In practice, we use the following formula, because it makes the calculations much quicker and easier:

$$\text{var} = \frac{\sum x^2}{n} - \left(\frac{\sum x}{n}\right)^2$$

Here is an example. Find the variance of the values 3, 7, 12.

$$\sum x = 3 + 7 + 12 = 22$$

$$\sum x^2 = 3^2 + 7^2 + 12^2 = 9 + 49 + 144 = 202$$

$$\text{var} = \frac{\sum x^2}{n} - \left(\frac{\sum x}{n}\right)^2 = \frac{202}{3} - \left(\frac{22}{3}\right)^2 = 67.33 - (7.33)^2 = 67.33 - 53.78 = 13.55$$

Hence:

standard deviation = $\sqrt{\text{variance}}$ = $\sqrt{13.55}$ = 3.68

This answer indicates that, on average, the original values are 3.68 away from the mean. (From above, the mean is 7.33.)

Calculating the standard deviation from a frequency table requires extra columns to be added to the table, as follows:

| x | f | Mid-points, x | fx | fx² |
|---|---|---|---|---|
| 1–10 | 15 | 5.5 | 82.5 | 453.75 |
| 11–20 | 39 | 15.5 | 604.5 | 9 369.75 |
| 21–30 | 62 | 25.5 | 1581.0 | 40 315.50 |
| 31–40 | 41 | 35.5 | 1455.5 | 51 670.25 |
| 41–50 | 22 | 45.5 | 1001.0 | 45 545.50 |
| | $\sum f = 179$ | | $\sum fx = 4724.5$ | $\sum fx^2 = 147354.75$ |

*Hint*: a common mistake that students make in completing the above table is to think that the values for the last column, labelled $fx^2$, are found by squaring the values in the column labelled $fx$. Note that $fx^2$ is not $(fx)^2$. To find $fx^2$, the easiest method is to multiply $x$ by $fx$.

The calculations for the mean and standard deviation from the table are as follows:

$$\text{mean, } \bar{x} = \frac{\sum fx}{\sum f} = \frac{4724.5}{179} = 26.4$$

$$\text{standard deviation} = \sqrt{\frac{\sum fx^2}{\sum f} - \left(\frac{\sum fx}{\sum f}\right)^2} = \sqrt{\frac{147354.75}{179} - \left(\frac{4724.5}{179}\right)^2}$$

$$= \sqrt{\left(823.21 - 26.394^2\right)} = \sqrt{126.568} = 11.25$$

## Coding

The above calculations are particularly messy and time-consuming because of the following:

- the decimal values for the mid-points of the intervals
- the relatively high values for the frequencies

It is possible to speed up the calculations for the mean and standard deviation by applying a simple **coding** to the data. This method of calculation was originally designed to enable the numerical work to be done without a calculator. Calculators are permitted in the examination, but occasionally a question may be set for which a coding method is required.

The example below shows how coding can be used. It is the same as the example above so the answers are already known.

- mean = 26.4
- standard deviation = 11.25

| x | f | Mid-points, x | $y = \dfrac{x - 25.5}{10}$ | fy | fy² |
|---|---|---|---|---|---|
| 1–10 | 15 | 5.5 | −2 | −30 | 60 |
| 11–20 | 39 | 15.5 | −1 | −39 | 39 |
| 21–30 | 62 | 25.5 | 0 | 0 | 0 |
| 31–40 | 41 | 35.5 | 1 | 41 | 41 |
| 41–50 | 22 | 45.5 | 2 | 44 | 88 |
| | $\sum f = 179$ | | | $\sum fy = 16$ | $\sum fy^2 = 228$ |

Here the coding is $y = \dfrac{x - 25.5}{10}$

There are many other possible codings, but:
- subtracting the midpoint of the biggest group and
- dividing by the class width (of that biggest group)

usually involves the easiest set of calculations.

In an exam question, you are likely to be *given* the coding to be used, because it makes the examiner's job much easier if everyone uses the same coding.

The mean and standard deviation for y can be found from the table.

$$\text{mean, } \bar{y} = \frac{\sum fy}{\sum f} = \frac{16}{179} = 0.0894$$

$$\text{standard deviation for } y, \sqrt{\frac{\sum fy^2}{\sum f} - \left(\frac{\sum fy}{\sum f}\right)^2} = \sqrt{\frac{228}{179} - \left(\frac{16}{179}\right)^2} = \sqrt{1.2658} = 1.125$$

The results for y must be decoded in order to find the mean and standard deviation for x.

The original coding was:

$$y = \frac{x - 25.5}{10}$$

Rearranging this formula gives:

$$x = 25.5 + 10y$$

This also means that:

$$\bar{x} = 25.5 + 10\bar{y}$$

Hence:

the mean, $\bar{x} = 25.5 + (10 \times 0.0894) = 25.5 + 0.894 = 26.4$ (to 3 s.f.)

To decode the standard deviation, the full equation $x = 25.5 + 10y$ is not needed. Only the last part is required, so the standard deviation for $x = 10 \times$ standard deviation for $y$. Hence, the standard deviation for $x = 10 \times 1.125 = 11.25$.

The reason why we do not add the 25.5 when decoding the standard deviation is that standard deviation measures the *spread* of the data. If we add or subtract any constant value to all the $x$ values, it simply moves them up or down the number line. It does not alter their spread. However, if we multiply or divide the values, the spread is altered.

You can see how much quicker the calculations are in this example if the coding is applied.

## Skewness

When you have to comment on the shape of a distribution, you need to assess the skewness of the data. There are several ways to determine the skewness of a distribution. The table below summarises the different kinds of skewness.

| | Positive skew | Symmetric | Negative skew |
|---|---|---|---|
| If a histogram has already been drawn, use its shape to determine the skewness. | | | |
| If the mode, median and mean have already been calculated, use these statements to determine the skewness. | mode < median < mean | mode = median = mean | mode > median > mean |
| If the quartiles have already been found, use these statements to determine the skewness. | $Q_2 - Q_1 < Q_3 - Q_2$ | $Q_2 - Q_1 = Q_3 - Q_2$ | $Q_2 - Q_1 > Q_3 - Q_2$ |
| If a box plot has already been drawn, use the box to determine the skewness. | | | |

The last two rows in the table are both based on the quartiles. Even in the box plot, it is the position of the median in relation to the quartiles (i.e. the box itself) that determines the nature of the skew.

There are several ways of calculating the skewness of a set of data. If you are asked to do this in an examination, *you will always be given a formula* to use.

One formula for calculating the value of the skewness is:

$$\text{skewness} = \frac{3(\text{mean} - \text{median})}{\text{standard deviation}}$$

For a given set of data, you may be asked whether it is better to use the mean and standard deviation or the median and interquartile range to represent the data.

- If the data are skewed, using the median and interquartile range is the preferred option. Skewed data must have some extreme values, which would seriously affect the mean.
- If the data are symmetric, the mean and standard deviation are preferred.

# Probability

There are three simple but important points to note about probability, which are all covered on GCSE specifications but are still relevant to this module.

**(1)** Probability is defined as:

$$\text{probability of an event occurring} = \frac{\text{number of favourable events}}{\text{number of possible events}}$$

**(2)** The probability of an event *not* occurring is (1 − probability it does occur).

The notation commonly used is:
- Probability that event A occurs is written as $P(A)$.
- Probability that event A does not occur is written as $P(A')$.
- $P(A') = 1 - P(A)$

**(3)** All probabilities lie within the range $0 \le P(A) \le 1$.

## Sample space or probability space

A sample space is a grid, diagram or list of all the possible outcomes in a probability experiment. If a probability space has been drawn, it is usually easy to count the number of favourable outcomes in order to answer a question.

For example, if two dice are rolled and the question is concerned with the total score on the two dice, it is recommended that you draw the following sample space.

| 6 | 7 | 8 | 9 | 10 | 11 | 12 |
|---|---|---|---|----|----|----|
| 5 | 6 | 7 | 8 | 9 | 10 | 11 |
| 4 | 5 | 6 | 7 | 8 | 9 | 10 |
| 3 | 4 | 5 | 6 | 7 | 8 | 9 |
| 2 | 3 | 4 | 5 | 6 | 7 | 8 |
| 1 | 2 | 3 | 4 | 5 | 6 | 7 |
|   | 1 | 2 | 3 | 4 | 5 | 6 |

**Score on 2nd dice** (vertical axis)

**Score on 1st dice**

From this diagram, it is easy to count how many totals are, for example, equal to 8 or more.

$$\text{probability (total of 8 or more)} = \frac{15}{36} = \frac{5}{12}$$

*Note*: all probabilities must be given as a fraction, decimal or percentage.

*Hint*: an answer given as a fraction is often better than a decimal because the fraction is an exact answer, whereas the decimal may be rounded off to a certain number of decimal places and hence is only an approximation. If a question asks for an *exact answer*, it is likely that a fractional answer is needed. For example, if the answer is 2/3, writing 0.67 is likely to lose 1 mark.

## Venn diagrams

Certain types of probability questions can be summarised by using a Venn diagram.

For example, in a sixth form of 113 students, 65 study mathematics, 50 study chemistry and 23 study both. If we label the two groups as M for mathematics students and C for chemistry students, the Venn diagram looks like this:

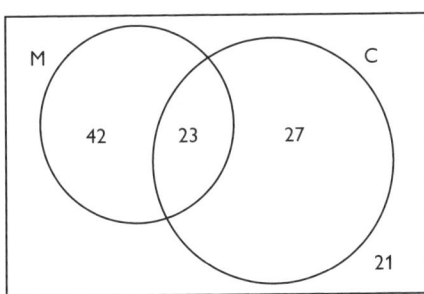

The information can be filled into the spaces as follows.
- 23 students study both subjects, so this value goes into the middle section, where the two loops overlap.

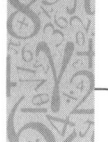

- There are 65 maths students altogether. We have already entered 23 of them in the central part, so the remaining 42 students are placed in the left-hand part of the M loop.
- Similarly, there are a further 27 chemistry students to put into the right-hand part of the C loop.
- Adding the values already entered, we have 42 + 23 + 27 = 92 students in the diagram. There are, according to the question, 113 students altogether, so 113 – 92 = 21 students must be put into the space outside the two loops. These are the students who study neither maths nor chemistry.

When you are constructing a Venn diagram, remember that:
- the box around the outside *must* be drawn — you will lose a mark in the examination if you leave out the box
- the number of items in the outer part of the diagram (in this case 21) *must be calculated and included* — this, too, will score a mark and the actual value may be required in a later part of the question

The Venn diagram can now be used to answer further questions. For example:
**(1)** Find the probability that a student chosen at random from the sixth form studies chemistry but not maths.

*Answer*: From the Venn diagram, it can be seen that there are 27 students inside the C loop but not inside the M loop. Hence the required probability is 27/113.

**(2)** Find the probability that a student chosen at random from the sixth form studies either maths or chemistry.

*Answer*: In any question asking for 'either...or...' you must always include the part that represents 'both'. In this case we need those who study maths but not chemistry (42 students), those who do chemistry but not maths (27 students) *and* those who do both maths and chemistry (23 students).

Altogether we have 42 + 27 + 23 = 92 students who do maths or chemistry or both:

$$\text{probability (maths or chemistry)} = \frac{92}{113}$$

Note that Venn diagrams can contain whole numbers (as above), percentages or probabilities.

The following notation is sometimes used in questions about Venn diagrams:
- $P(M \cap C)$ means the probability of M and C both occurring. The symbol '$\cap$' is known as the **intersection** of the two sets M and C. On the Venn diagram above, this is the overlapping section that is in both M and C (containing the value 23 above).
- $P(M \cup C)$ means the probability of M or C occurring. The symbol '$\cup$' is the **union** of sets M and C. On the Venn diagram above, this is made up of the three sections that are in either M or C or both (containing the values 42, 27 and 23 above).

## The addition rule

Using this new notation and with reference to the following diagram, in which each of the sections are numbered, we can state one important rule in probability.

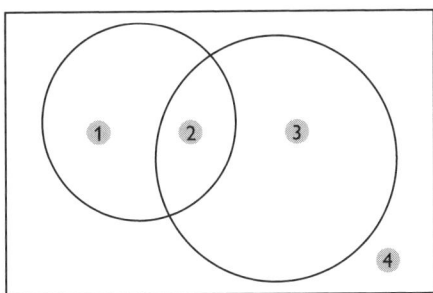

The addition rule states that:

$P(A \cup B) = P(A) + P(B) - P(A \cap B)$

That is, the probability of A or B is equal to the probability of A plus the probability of B minus the probability of both.

Using the diagram above, $P(A)$ is covered by sections 1 and 2; $P(B)$ is covered by sections 2 and 3. Therefore, $P(A) + P(B)$ means that sections 1 and 2 and sections 2 and 3 are included in the calculation, so section 2 in the diagram has been included twice. Hence, we need to subtract $P(A \cap B)$ to get rid of one occurrence of section 2.

The addition rule is widely used in many aspects of probability.

## Conditional probability

Looking back at the first Venn diagram, concerning the students in a sixth form (p. 27), the probability that a student selected at random studies chemistry is 50/113.

Suppose we now go into the maths classroom and select a student at random. The probability that this student studies chemistry is equal to 23/65.

This is because we already know that the student is one of the 65 maths students, and 23 maths students also study chemistry (the other 42 study only maths). This is a **conditional probability**.

We have found the probability that the student studies chemistry *given that* he/she studies maths. The notation used for this kind of probability is $P(C \mid M)$. The vertical line is the notation used to represent 'given that'.

We can now say that: $\qquad P(C \mid M) = \dfrac{23}{65}$

This could be written as: $\qquad P(C \mid M) = \dfrac{P(C \cap M)}{P(M)}$

From the Venn diagram: $\qquad P(C \cap M) = \dfrac{23}{113}$ and $P(M) = \dfrac{65}{113}$

Hence:

$$\frac{P(C \cap M)}{P(M)} = \frac{\frac{23}{113}}{\frac{65}{113}}$$

which simplifies to 23/65, as shown on p. 29.

The conditional law states that for two events A and B:

$$P(B|A) = \frac{P(A \cap B)}{P(A)} \quad \text{or} \quad P(A|B) = \frac{P(A \cap B)}{P(B)}$$

In particular, note that the denominator is always the 'given that' event.

## The multiplication rule

The conditional law:

$$P(B|A) = \frac{P(A \cap B)}{P(A)}$$

can be rearranged to give the **multiplication rule**, which states that:

$$P(A \cap B) = P(A) \times P(B|A)$$

## Tree diagrams

Tree diagrams can be used in probability for any problem that involves two or more stages — for example, choosing two or more counters from a box. Tree diagrams are another way of using a diagram to illustrate and clarify a situation.

Imagine a box containing five black counters and three red counters. Two counters are selected at random from the box. This is an example of a situation that is **without replacement**. In other words, we assume that the two counters are chosen simultaneously. When selecting the first counter, there are eight to choose from, but when selecting the second counter, there are only seven left in the bag.

The tree diagram for this situation would look like this:

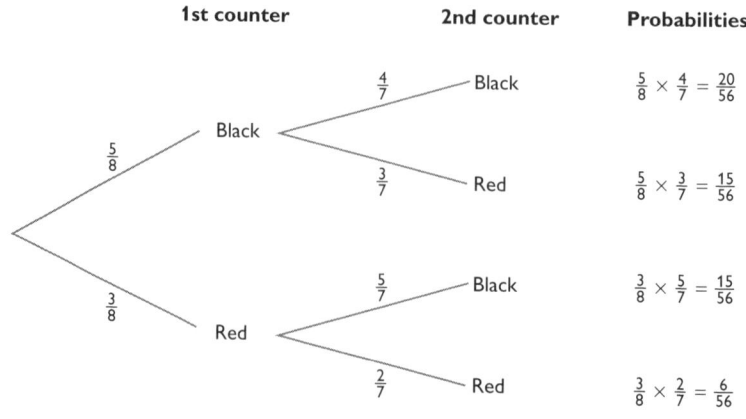

The probability of both counters being black is given by:

$$\frac{5}{8} \times \frac{4}{7} = \frac{20}{56}$$

Starting at the left-hand side of the diagram, we *multiply* the probabilities as we move along the branches of the tree.

To find the probability of getting one red and one black counter, we need to look at two of the branch endings, because the required outcome could be either (black, red), with probability 15/56 or (red, black) with probability 15/56. Adding these two probabilities gives an answer of 30/56 or 15/28.

*Note*: if the first counter is returned to the box *before* the second counter is selected, this is '**with replacement**' and the probabilities for the second selection are the same as for the first.

## Mutually exclusive events

A and B are mutually exclusive events if $P(A \cap B) = 0$. In other words the two events cannot possibly happen together. For example, imagine we pick a card from an ordinary 52-card pack:

- Event A is getting a diamond.
- Event B is getting a black card.

It is impossible for both A and B to occur simultaneously, so A and B are mutually exclusive events.

In an exam question, you may be asked to decide whether A and B are mutually exclusive. This requires you to show, from your previous calculations and results, whether $P(A \cap B) = 0$.

If A and B are mutually exclusive, the two loops on the Venn diagram do not overlap and the Venn diagram looks like this:

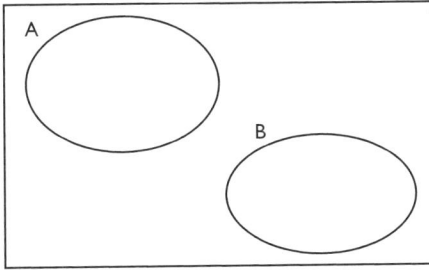

## Independent events

When A and B are independent, if A occurs then it does not alter the probability of B occurring, and vice versa. This means that $P(B \mid A) = P(B)$, indicating that the probability of B occurring is exactly the same, whether A happens or not.

For example, if you are finding the probability of scoring a 6 when rolling a dice and getting a head when tossing a coin, if you have already noted a score of 6 on the dice then the probability of a head on the coin is still 1/2. The two events are independent.

If A and B are independent, the multiplication rule can be simplified from:

$$P(A \cap B) = P(A) \times P(B \mid A)$$

to

$$P(A \cap B) = P(A) \times P(B).$$

This simplified rule works only if A and B are independent.

In an exam question you may be asked to state whether A and B are independent. In order to do this it is necessary to use your previous calculations and answers to show that:

• either $P(A \cap B) = P(A) \times P(B)$
• or $P(A \mid B) = P(A)$; here $P(B \mid A) = P(B)$ would be an alternative solution

Either of the above methods could be used to determine whether A and B are independent. The method selected depends on what information has already been found in earlier parts of the problem.

# Correlation and regression

## Correlation

**Bivariate data** occur if we are investigating relationships between pairs of observations — for example, a possible relationship between weight and blood pressure. This would involve recording these two variables for a number of different individuals.

Bivariate data are represented by a **scatter diagram**. There are three possible patterns for a scatter diagram.

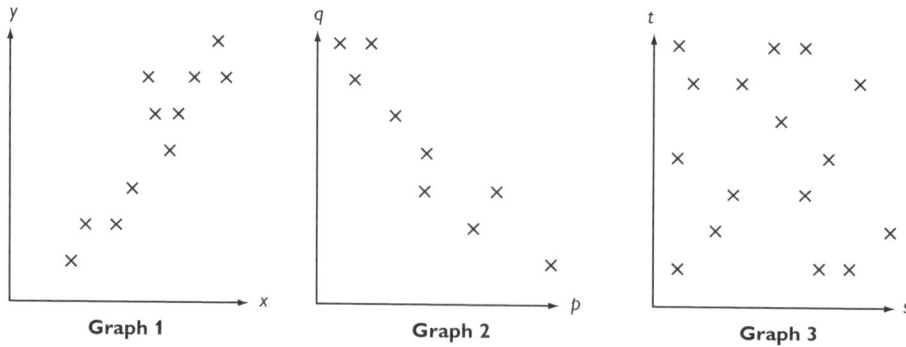

Graph 1          Graph 2          Graph 3

- Graph 1 shows **positive correlation**, because $y$ increases as $x$ increases.
- Graph 2 shows **negative correlation**, because $q$ decreases as $p$ increases.
- Graph 3 shows **no correlation**, because the points are scattered randomly.

The **product–moment correlation coefficient (PMCC)** gives a numerical value to indicate how strong a possible link between two variables might be.

For the two variables $x$ and $y$, the product–moment correlation coefficient is given by:

$$r = \frac{S_{xy}}{\sqrt{\left(S_{xx}S_{yy}\right)}}$$

This formula appears in the Edexcel exam formula book and for Unit 6683 you do not need to know the underlying theory. All that is required is the ability to:

- perform the necessary calculations, with the aid of the given formulae
- interpret the resulting correlation value

The formulae for $S_{xx}$, $S_{yy}$ and $S_{xy}$ are also included in the formula book, but they can be stated as:

$$S_{xx} = \sum x^2 - \frac{\left(\sum x\right)^2}{n}, \; S_{yy} = \sum y^2 - \frac{\left(\sum y\right)^2}{n}, \; S_{xy} = \sum xy - \frac{\left(\sum x \sum y\right)}{n}$$

The following example shows how these formulae work.

The data in the table represent the age, $x$, in weeks and the weight, $y$, in kilograms of eight babies weighed at a clinic.

| Age, $x$/weeks | Weight, $y$/kg | $x^2$ | $y^2$ | $xy$ |
|---|---|---|---|---|
| 2 | 3.3 | 4 | 10.89 | 6.6 |
| 5 | 4.2 | 25 | 17.64 | 21.0 |
| 8 | 4.7 | 64 | 22.09 | 37.6 |
| 9 | 5.0 | 81 | 25.00 | 45.0 |
| 9 | 4.3 | 81 | 18.49 | 38.7 |
| 11 | 4.9 | 121 | 24.01 | 53.9 |
| 12 | 5.8 | 144 | 33.64 | 69.6 |
| 15 | 6.1 | 225 | 37.21 | 91.5 |
| $\sum x = 71$ | $\sum y = 38.3$ | $\sum x^2 = 745$ | $\sum y^2 = 188.97$ | $\sum xy = 363.9$ |

You may be required to draw a scatter diagram to illustrate the given data. The scatter diagram for these data is shown on p. 34.

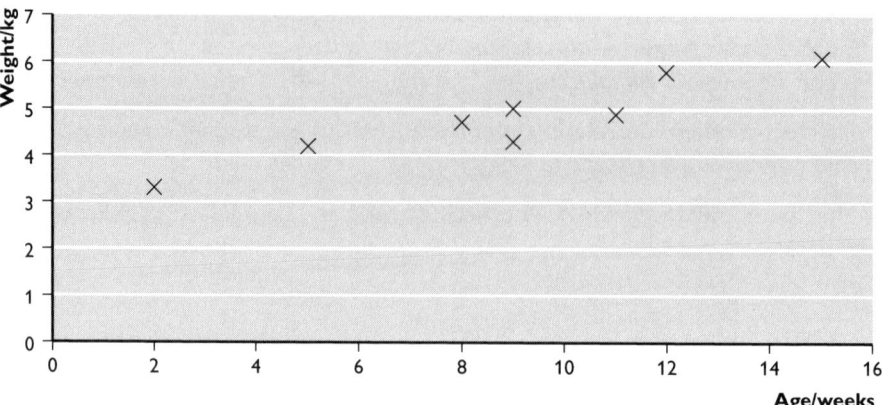

Often the first part of an exam question asks you to calculate the values of $S_{xx}$, $S_{yy}$ and $S_{xy}$. You would then go on to use these values in the formula for the correlation coefficient.

$$S_{xx} = \sum x^2 - \frac{(\sum x)^2}{n} = 745 - \frac{71^2}{8} = 114.875$$

$$S_{yy} = \sum y^2 - \frac{(\sum y)^2}{n} = 188.97 - \frac{38.3^2}{8} = 5.60875$$

$$S_{xy} = \sum xy - \frac{(\sum x \sum y)}{n} = 363.9 - \frac{71 \times 38.3}{8} = 23.9875$$

$$r = \frac{S_{xy}}{\sqrt{(S_{xx} S_{yy})}} = \frac{23.9875}{\sqrt{(114.875 \times 5.60875)}} = 0.945$$

The correlation value should be expressed correct to 3 significant figures, unless the question ask for a different level of accuracy. However, you should not round off the values of $S_{xx}$, $S_{yy}$ or $S_{xy}$ to 3 significant figures. This would lead to inaccuracies in the later calculations and consequently a loss of accuracy marks.

Once the calculation has been completed, the value of the correlation coefficient can be interpreted. A correlation coefficient of $r = 0.945$ shows a strong positive correlation. This suggests that the older the baby, the heavier it will be. It is important to remember that the interpretation must be in context. It is true to state that it is a strong positive correlation, but the proper interpretation is a statement such as 'the older the baby, the heavier it will be'. It is this part of the interpretation that scores the mark in an examination.

Important note: $-1 \le r \le +1$
- A strong correlation is shown by values that are close to either 1 or −1.
- For example, a correlation coefficient of −0.75 suggests a strong negative correlation.
- A value close to zero (positive or negative) means there is little or no correlation.

# Regression

To find a rule that connects two variables, we need to look for a **regression equation**. This is sometimes referred to as the **line of best fit** on a scatter diagram.

The particular method required in this unit is the **least-squares regression line**. As with the formulae in the section on correlation, the theory for the least-squares regression line is beyond the requirements of the unit and all the formulae appear in the Edexcel examination booklet.

The main purpose of a regression equation is that it can be used to *predict* a value of the **response variable** for a given value of the **explanatory variable**. On a scatter diagram the explanatory variable is usually drawn as the horizontal axis, with the response variable on the vertical axis.

The equation for the variables $x$ and $y$ is known as the **equation of y on x** and takes the form:

$$y = a + bx$$

In this equation, the values of $a$ and $b$ must be calculated using the formulae:

$$b = \frac{S_{xy}}{S_{xx}} \text{ and } a = \bar{y} - b\bar{x}$$

Using the data from the example in the correlation section above, we might want to estimate the weight of a baby of a given age, in weeks. We would need the regression equation of $y$ on $x$.

We have already calculated the following values:

$$S_{xx} = 114.875 \text{ and } S_{xy} = 23.9875$$

So:

$$b = \frac{S_{xy}}{S_{xx}} = \frac{23.9875}{114.875} = 0.2088139$$

and

$$a = \bar{y} - b\bar{x} = \frac{38.3}{8} - 0.2088139 \times \frac{71}{8} = 2.934277$$

The least-squares regression equation of $y$ on $x$ is:

$$y = 2.93 + 0.209x$$

As with the correlation coefficient, the values of $a$ and $b$ should be given correct to 3 significant figures, unless the question states otherwise.

There are three types of question that often follow on from the regression equation. For example:

**(1) Estimate the weight of a 10-week-old baby.**

Here we substitute into the equation:

when $x = 10$ weeks, $y = 2.93 + 0.209 \times 10 = 5.02$ kg

**(2) Interpret the values of $a$ and $b$ in your equation.**

The interpretations must be stated in the context of the question. Stating that $a$ is the intercept on the $y$-axis would not score the mark. The correct interpretation of $a$ is that 2.93 kg in this example is the expected weight of a newborn baby (i.e. the weight when a baby is zero weeks old).

Similarly, writing just that $b$ is the gradient of the line would score nothing. The interpretation of $b$ needs to explain what this rate of change actually means. The value 0.209 is the increase in a baby's weight for each additional week.

**(3) For what range of values of $x$ are the estimates from the regression equation likely to be reliable?**

We can use the equation to interpolate, in other words to make estimates within the range of the collected data.

In this example, the babies were aged between 2 and 15 weeks, so the range would be $2 \leq x \leq 15$.

We do not know whether the linear model would continue to be appropriate for values above 15 weeks, so extrapolation (reading outside the range) is not appropriate and would provide unreliable estimates.

## Coding

The concept of coding has already been mentioned as a means of simplifying data values (see pp. 23–25). This same principle can be used in both correlation and regression.

If we consider the example concerning the age and weight of eight babies, we can use the following simple codes:

$s = x - 2$ and $t = 10y - 30$

This gives values of $s$ of 0, 3, 6, 7,... etc. and values of $t$ of 3, 12, 17, 20,... etc.

The summary values are then:

$$\sum s = 55, \ \sum t = 143, \ \sum s^2 = 493, \ \sum t^2 = 3117 \text{ and } \sum st = 1223$$

and hence $S_{ss} = 114.875$, $S_{tt} = 560.875$ and $S_{st} = 239.875$.

This, in turn, gives a correlation coefficient of $r = 0.945$. Note that this is exactly the same as the value before the coding. No decoding is necessary for the correlation coefficient.

For the regression equation of $t$ on $s$, the summary values above give:

$b = 2.08814$ and $a = 3.51904$

Hence, the regression equation of $t$ on $s$ is:

$t = 3.52 + 2.088s$

To decode this expression, so that it becomes the equation of $y$ on $x$, we insert the two codes into the equation for $t$ on $s$:

$t = 3.52 + 2.088s$

- Putting $t = 10y - 30$ and $s = x - 2$, we get $(10y - 30) = 3.52 + 2.088(x - 2)$
- Removing the brackets, $10y - 30 = 3.52 + 2.088x - 4.176$
- Rearranging, $10y = 30 + 3.52 + 2.088x - 4.176$
- Simplifying, $10y = 29.344 + 2.088x$
- Dividing by 10, $y = 2.93 + 0.209x$

This is the $y$ on $x$ equation that we found earlier (see p. 35).

# Discrete random variables

Discrete random variables are discrete variables that change value according to some probability rule.

The notation used throughout this section is as follows:
- Upper-case letters (capitals) indicate the label or name of a variable.
- Lower-case letters (small letters) indicate the values the variable can take.

So we could say:
- $X$ is the variable 'total score when two dice are rolled'
- $x = 2, 3, 4, 5, 6, 7, 8, 9, 10, 11, 12$

## Probability functions

The probability function of a random variable can be described in two ways. The first way is to use a table of probabilities, as shown in the example below.

| $x$ | 0 | 1 | 2 | 3 |
|---|---|---|---|---|
| $P(X = x)$ | 0.5 | 0.3 | 0.1 | 0.1 |

The first column of data states that the probability that $x = 0$ is 0.5.

The table of probabilities is known as the **probability distribution** of $x$. In such a table, the probabilities must add up to 1.0, because the table represents all the possible outcomes. It is certain that one of them will occur if the probability experiment takes place.

The second way to describe a probability function is in the form:

$$P(X=x) = \begin{cases} \dfrac{1}{100}x^3 & x = 1, 2, 3, 4 \\ 0 & \text{otherwise} \end{cases}$$

The first line of the function gives a rule for finding the probabilities associated with each of the four values of $x$. The second line simply states that the probability is equal to zero for all other values of $x$.

If a probability function is given in this form, it is usually easier to proceed with the question by changing it into a table of probabilities first. So the function given above would look like this in tabular form:

| $x$ | 1 | 2 | 3 | 4 |
|---|---|---|---|---|
| $P(X = x)$ | $\dfrac{1}{100}$ | $\dfrac{8}{100}$ | $\dfrac{27}{100}$ | $\dfrac{64}{100}$ |

# Cumulative distribution function

Definition: $F(x) = P(X \leq x)$

This is a **cumulative probability**. It is the probability that $x$ is less than or equal to a given value.

In the table above, $F(3) = P(X \leq 3) = P(X = 1) + P(X = 2) + P(X = 3)$

$$= \dfrac{1}{100} + \dfrac{8}{100} + \dfrac{27}{100} = \dfrac{36}{100} = \dfrac{9}{25}$$

*Note:* $F(2.7) = P(X \leq 2.7)$

In this example this is the same as $P(X \leq 2)$ because there are no values in the distribution that are not whole numbers.

Similarly, $F(25) = P(X \leq 25)$. In the example the highest possible value for $x$ is 4, so it is certain that $x \leq 25$. Hence $F(25) = 1$.

# Expectation and variance

## Expectation (or mean)

The **expectation** of the variable $X$ is written as $E(X)$. It can also be referred to as the **expected value** of $X$ or the **mean** of $X$.

In the earlier section concerning means from a frequency distribution (p. 21), the formula given for the mean was:

$$\text{mean} = \dfrac{\sum fx}{\sum f}$$

For probability distributions, rather than frequency distributions, we replace the frequencies, $f$, with probabilities, $p$. The formula for the mean or for $E(X)$ becomes:

$$\text{mean} = E(X) = \frac{\sum px}{\sum p}$$

Since the denominator of the fraction is $\sum p = 1$, because it represents total probability, the formula can be written as:

$$E(X) = \sum px$$

In this formula, $p$ is the shorthand for $P(X = x)$, so, when written out in full, the formula becomes:

$$E(X) = \sum xP(X = x)$$

For example, if the probability distribution is:

| $x$ | 1 | 2 | 3 |
|---|---|---|---|
| $P(X = x)$ | 0.2 | 0.3 | 0.5 |

$$E(X) = (1 \times 0.2) + (2 \times 0.3) + (3 \times 0.5) = 0.2 + 0.6 + 1.5 = 2.3$$

## Variance

The variance of a probability distribution is written as $\text{Var}(X)$.

In the earlier section on standard deviation and variance (pp. 22–25), the formula used was:

$$\text{Var} = \frac{\sum fx^2}{\sum f} - \left(\frac{\sum fx}{\sum f}\right)^2$$

Replacing the frequencies, $f$, with probabilities, $p$, as in the formula for $E(X)$ above, gives a formula for $\text{Var}(X)$:

$$\text{Var}(X) = \sum px^2 - \left(\sum px\right)^2$$

When written out in full, this becomes:

$$\text{Var}(X) = \sum x^2P(X = x) - \left(\sum xP(X = x)\right)^2$$

or

$$\text{Var}(X) = \sum x^2P(X = x) - \text{mean}^2$$

From the probability distribution in the example above:

$$\text{Var}(X) = \left(1^2 \times 0.2\right) + \left(2^2 \times 0.3\right) + \left(3^2 \times 0.5\right) - 2.3^2$$

$$= 0.2 + 1.2 + 4.5 - 2.3^2 = 0.61$$

*Note:* it is easier to remember and use the two shorter versions of the formula:

$$E(X) = \sum px \text{ and } \text{Var}(X) = \sum px^2 - \left(\sum px\right)^2$$

An alternative version of the formula for $\text{Var}(X)$ is often seen and is needed to solve certain kinds of problem:

$$\text{Var}(X) = E(X^2) - \left(E(X)\right)^2$$

This is equivalent to the formula used on p. 39, but it is written out in terms of expected values, rather than using sigma notation.

Sometimes an examination question may specifically ask you to find $E(X^2)$.

If $E(X) = \sum xP(X = x)$, then $E(X^2) = \sum x^2 P(X = x)$

# Algebra of expectation and variance

If the values of $E(X)$ and Var($X$) have already been found, it is possible to use the results to find other expected values or variances of different functions of the variable $X$.

The important rules are as follows, where $a$ and $b$ are constants:
- $E(aX) = aE(X)$
- $E(aX + b) = aE(X) + b$
- $E(aX - b) = aE(X) - b$
- $E(aX + bY) = aE(X) + bE(Y)$
- $E(aX - bY) = aE(X) - bE(Y)$
- $\text{Var}(aX) = a^2\text{Var}(X)$
- $\text{Var}(aX + b) = a^2\text{Var}(X)$
- $\text{Var}(aX - b) = a^2\text{Var}(X)$
- $\text{Var}(aX + bY) = a^2\text{Var}(X) + b^2\text{Var}(Y)$
- $\text{Var}(aX - bY) = a^2\text{Var}(X) + b^2\text{Var}(Y)$

Notice that the final two expressions are identical, whether we are adding or subtracting the variances.

For example, if $E(X) = 6$, Var($X$) = 2, $E(Y) = 8$ and Var($Y$) = 3:
- $E(3X + 5Y) = 3E(X) + 5E(Y) = 3 \times 6 + 5 \times 8 = 18 + 40 = 58$
- $E(3X - 5Y) = 3E(X) - 5E(Y) = 3 \times 6 - 5 \times 8 = 18 - 40 = -22$
- $\text{Var}(3X + 5Y) = 3^2\text{Var}(X) + 5^2\text{Var}(Y) = 9 \times 2 + 25 \times 3 = 18 + 75 = 93$
- $\text{Var}(3X - 5Y) = 3^2\text{Var}(X) + 5^2\text{Var}(Y) = 93$ (as above)
- $E(7 - 2Y) = 7 - 2E(Y) = 7 - 2 \times 8 = 7 - 16 = -9$
- $\text{Var}(7 - 2Y) = 2^2\text{Var}(Y) = 4 \times 3 = 12$

# The discrete uniform distribution

Consider the probability distribution for the number $X$ that occurs when a single dice is rolled:

| x | 1 | 2 | 3 | 4 | 5 | 6 |
|---|---|---|---|---|---|---|
| P(X = x) | $\frac{1}{6}$ | $\frac{1}{6}$ | $\frac{1}{6}$ | $\frac{1}{6}$ | $\frac{1}{6}$ | $\frac{1}{6}$ |

This is an example of the discrete uniform distribution. The key features of this distribution are that:
- the values of $x$ are all discrete, starting at 1 and going up to $n$
- all the probabilities are equal

If you are asked to name this distribution in an examination question, you must ensure that both the words 'discrete' and 'uniform' are included.

The probability distribution can be written out as a table of values or it can be expressed as a probability function. The probability function for the rolling of a single dice is:

$$P(X=x) = \frac{1}{6}, \quad x = 1, 2, 3, 4, 5, 6$$

## Expectation and variance of a discrete uniform distribution

From the table of values on p. 40, it is possible to use the formulae:

$$E(X) = \sum px \text{ and } Var(X) = \sum px^2 - \left(\sum px\right)^2$$

from the previous section in order to calculate $E(X)$ and $Var(X)$.

Alternatively, it is much quicker if you learn the following formulae, which do not appear in the examination formula booklet.

For a discrete uniform distribution:

$$\text{mean} = E(X) = \frac{1+n}{2}, \ Var(X) = \frac{n^2 - 1}{12}$$

So, if we were rolling a fair eight-sided dice:

$$x = 1, 2, 3,..., 7, 8 \text{ and } p = \frac{1}{8}$$

This gives $E(X) = \dfrac{1+8}{2} = 4.5$ and $Var(X) = \dfrac{8^2 - 1}{12} = \dfrac{63}{12} = 5.25$

# The normal distribution

The **normal distribution** fits a large proportion of data that are *continuous*, such as height, weight, time, age etc. It is the most important distribution in statistics. It is used extensively in further applications of statistics that go beyond the scope of Unit 6683.

The following sketch shows the general shape of the normal distribution:

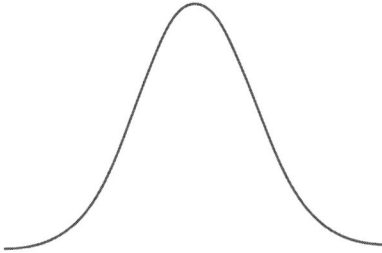

If a histogram is drawn to represent a set of data and the histogram is the same basic shape as this curve, it is highly likely that the data would be modelled by the normal distribution.

In an examination question, you may be asked to list the following key features of a normal distribution:
- It is symmetrical about the mean.
- mode = median = mean
- It has a bell-shaped curve (see p. 41).
- The horizontal axis is an asymptote to the curve.
- 95% of the data lie within 2 standard deviations of the mean (or 68% lie within 1 standard deviation or almost all the data lie within 3 standard deviations of the mean).

Consider a set of data for the heights (in cm) of a large number of adult males and another set of data for the time taken (in seconds) to change a large number of car tyres. The histograms representing the two sets of data would obviously be different. They would have different means and different standard deviations. However, it is likely that the two histograms would both fit the symmetric, bell-shaped curve. In other words, they are both likely to fit the conditions for the normal distribution.

## Notation

The notation $X \sim N(\mu, \sigma^2)$ is commonly used to define a normal distribution.
- Upper-case $X$ is the label for the distribution.
- The symbol '$\sim$' stands for the phrase 'is distributed'.
- 'N' stands for normal distribution.
- $\mu$ is the mean and $\sigma^2$ is the variance of the distribution. These are the two **parameters** of the distribution.

## Standardising

We can use the mean, $\mu$, and standard deviation, $\sigma$, to standardise any values from a set of data that are normally distributed. Standardising the data converts them from $x$ values into $z$ values, so:

$$X \sim N(\mu, \sigma^2) \text{ becomes } Z \sim N(0, 1)$$

This process of standardisation changes all normally distributed data into the same form where mean = 0 and standard deviation = variance = 1.

The standardisation is carried out using the formula:
$$z = \frac{X - \mu}{\sigma}$$

In turn, exactly the same method can be applied for estimating proportions and probabilities using prepared normal distribution tables.

The mathematical theory behind the equation of a normal distribution curve and the calculus required to find the associated areas under the curve are not required in this unit, but detailed explanations can be found in any statistics textbook.

Working through an example is the best way of illustrating how the various steps work in a calculation.

**A company produces light bulbs with a mean lifetime of 90 hours and a standard deviation of 6 hours. If a bulb is selected at random from a day's production, calculate the probability that it lasts longer than 100 hours. It can be assumed that the lifetimes of the bulbs follow a normal distribution.**

Here the variable can be defined as $X \sim N(90, 6^2)$. You need to calculate $P(X > 100)$, Which is equivalent to finding the shaded area on the following curve:

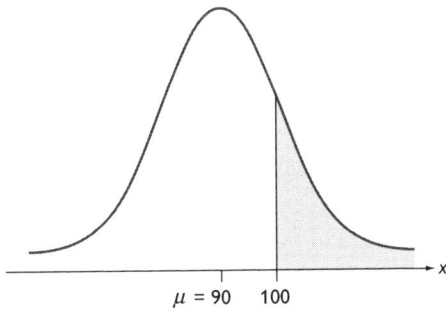

**Step 1**: standardise the variable.

$$Z = \frac{X - \mu}{\sigma}$$

We need to standardise the value, $x = 100$:

$$Z = \frac{100 - 90}{6} = 1.67$$

*Note*: the value of $z$ has been written correct to 2 decimal places, because the values of $z$ throughout the normal distribution tables are all correct to 2 decimal places.

**Step 2**: draw a sketch of the curve.

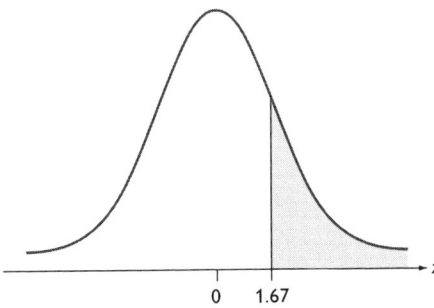

*Hint*: it is *strongly recommended* that you draw a sketch graph like the one above for *every* calculation using the normal distribution.

The area under the curve represents the probability.

In the normal distribution tables, looking up any $z$ value gives $P(Z < z)$, correct to 4 decimal places. In the tables this probability is labelled as $\Phi(z)$. $\Phi(z)$ is simply an alternative notation to $P(Z < z)$. This is always the area on the graph *to the left* of the $z$ value, as shown in the sketch on p. 44.

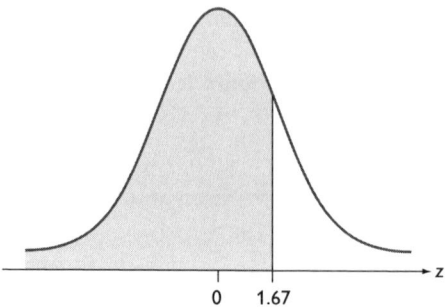

The total area under the curve represents total probability, so the total area is equal to 1. The curve is symmetrical, so we can find the areas to the right or left of a given $z$ value using this symmetry.

**Step 3**: use the normal distribution tables to find the required probability.

We want $P(Z > 1.67)$, which is the area to the right of $z = 1.67$:

$$P(Z > 1.67) = 1 - P(Z < 1.67)$$
$$= 1 - 0.9525 \text{ (from the table)}$$
$$= 0.0475$$

So, the probability that a bulb lasts for more than 100 hours is 0.0475.

## Negative $z$ values

There are no negative $z$ values in the normal distribution tables, but negative $z$ values often occur as a result of standardising the variable. Again, we can use the symmetry of the curve to deal with negative $z$ values.

For example, finding $P(Z > -1.85)$ is exactly the same as finding $P(Z < +1.85)$, as illustrated by the sketches below:

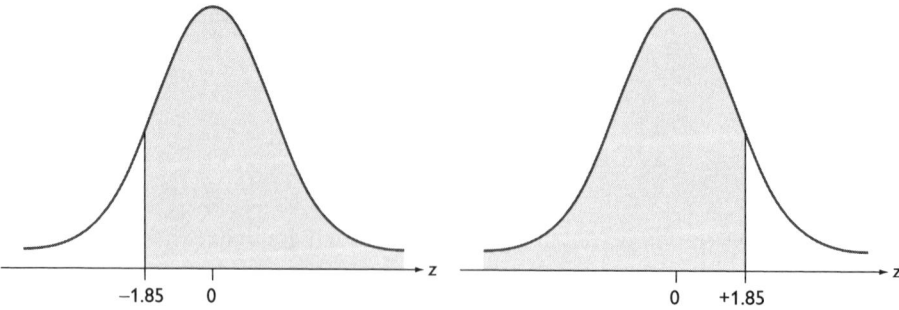

Hence, $P(Z > -1.85) = P(Z < +1.85) = 0.9678$

Similarly, $P(Z < -1.85) = P(Z > +1.85) = 1 - P(Z < +1.85) = 1 - 0.9678 = 0.0322$.

Often we need to find a probability that lies between two values of $z$. For example, $P(-1.4 < Z < 2.1)$ is represented by the shaded area in the curve on p. 45.

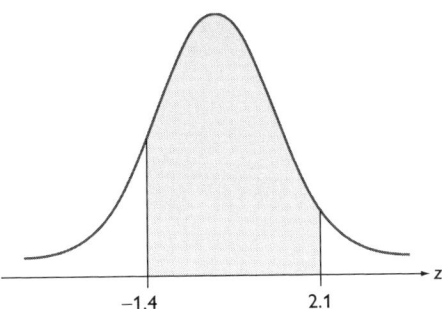

There are several ways of finding the required probability, but the most straightforward is to find the area to the left of both $z$ values and then calculate the difference between them. This can be written as:

$P(-1.4 < Z < 2.1) = P(Z < 2.1) - P(Z < -1.4)$

$P(Z < 2.1) = 0.9821$

$P(Z < -1.4) = 1 - P(Z < +1.4) = 1 - 0.9192 = 0.0808$

Hence, $P(Z < 2.1) - P(Z < -1.4) = 0.9821 - 0.0808 = 0.9013$

Exam questions on the normal distribution fall into two standard types:

**(1)** Finding a probability, when you are given the mean, $\mu$, and the standard deviation, $\sigma$ (or variance, $\sigma^2$). This type of question is illustrated by the earlier example concerning the lifetimes of light bulbs.

**(2)** Finding the mean and standard deviation (or variance), when you are given some alternative information about the distribution. This is illustrated in the next section.

## Finding $\mu$ and $\sigma^2$ in a normal distribution

**The random variable $X$ is normally distributed and it is known that $P(X > 76) = 0.15$ and $P(X < 63) = 0.0107$. Calculate the value of $\mu$ and the value of $\sigma$.**

It is helpful to sketch curves to represent the two pieces of information given in this question.

**(1) Considering $P(X > 76) = 0.15$**

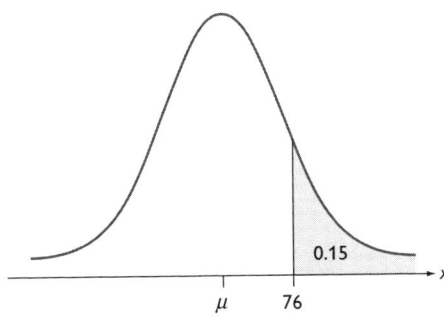

We need to find the value of $z$ that corresponds to a probability of 0.15.

The smaller table ('Percentage points of the normal distribution') should be checked first. In this table, the value $p = 0.1500$ can be seen, which gives a corresponding $z$ value of 1.0364.

We can use this value, together with the usual standardisation formula, $z = \dfrac{x - \mu}{\sigma}$, to construct the equation:

$$1.0364 = \frac{76 - \mu}{\sigma}$$

and this can be rearranged into:

$$\mu + 1.0364(\sigma) = 76 \quad \text{(equation 1)}$$

**(2) Considering $P(X < 63) = 0.0107$**

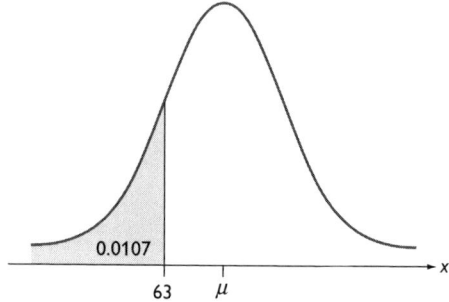

Again, we start by looking in the 'Percentage points of the normal distribution' table, but the probability 0.0107 does not appear in this table. This means we have to go back to the larger table and look for the value 0.0107 in the probabilities column, the column labelled as $\Phi(z)$. In this table, all the values of $\Phi(z)$ are equal to 0.5 or more.

We need to use the symmetry ideas from earlier to help find the $z$ value that corresponds to $\Phi(z) = 0.0107$. We simply calculate the area to the right of $x = 63$ on the curve, which is:

$$1 - 0.0107 = 0.9893$$

So if $\Phi(z) = 0.9893$, then $z = 2.30$.

Remember, again using the idea of symmetry, that $z = 2.30$ is on the right-hand side of the curve. Our diagram shows clearly that we need a $z$ value on the left-hand side. Hence $z = -2.30$.

This value of $z$ is now used to form the equation:

$$-2.30 = \frac{63 - \mu}{\sigma}$$

Rearranging gives $\mu - 2.30\sigma = 63$ (equation 2)

We now have two equations with two unknowns, $\mu$ and $\sigma$. These must be solved simultaneously to find the values of $\mu$ and $\sigma$.

$\mu + 1.0364\sigma = 76$  (equation 1)

$\mu - 2.30\sigma = 63$  (equation 2)

Subtracting gives $3.3364\sigma = 13$

Dividing gives $\sigma = 3.8964$

Substituting into equation 1:
$$\mu + (1.0364 \times 3.8964) = 76$$
$$\mu = 71.96$$

Hence we now have the values:
$\mu = 72.0$ and $\sigma = 3.90$ (both correct to 3 significant figures)

There are some important points to note in this type of question:
- Draw sketches of the curve to help avoid errors with the $z$ values.
- If possible, use the table of percentage points of the normal distribution to find a $z$ value corresponding to a given probability.
- If the probability does appear on the 'percentage points' table, make sure that you write down the $z$ value using all 4 decimal places.
- Remember that $z$ values on the left-hand side must be negative. (Forgetting this point is a common error in examination questions and it loses most of the marks for the question concerned.)

The following questions are designed to cover all the topics in Edexcel Unit 6683: Statistics 1. Different slants on each question have been included so that you get a feel for all types of question that could occur on a given topic.

For each question there is a written solution. Remember that the method you have learned may not be the same as the one offered in the solution. It is up to you to make decisions about whether you stick with your method or adapt some of the methods presented here.

## Mark schemes

A realistic mark scheme has been provided for each question so that you can judge for yourself how many marks you might lose for a solution that is not completely correct. The mark scheme works as follows.

- **M marks**: these are awarded for using the correct method. It is important to note that simply writing down a correct formula is not enough to gain a method mark. You have to attempt to *use* the formula by putting some numbers into it.
- **A marks**: these are for accuracy. You can score the A marks only if you gained the M mark first. If the method is wrong, you would get M = 0 and then you would get A = 0 automatically, even if, by some chance, you managed to obtain a correct answer.
- **B marks**: these are marks for answers that do not depend on a method. They are used for questions that require you to make a comment and for answers that are worth only a single mark.

## Examiner's comments

Interspersed with the solutions are examiner's comments, denoted by the symbol 🄴. In some cases the comments tell you the most common mistakes that are made on that type of question and give some hints on how to avoid making errors. In other cases the comments point out the main differences between A-grade and C-grade solutions.

## Question 1

**(a) Give three reasons for using a statistical model.** (3 marks)

**(b) Write down the name of the distribution you would suggest as a suitable model for each of the following situations:**

**(i) the time taken to complete a crossword**

**(ii) the number on the upper face of a fair dice** (2 marks)

## Answer to Question 1

**(a)** It simplifies a real-world problem.

It allows us to analyse or describe a real-world problem.

It is quicker and cheaper than using the real-life situation.

It is possible to predict future outcomes.

It is possible to refine a model. [Any 3 from this list; B1, B1, B1]

**(b)(i)** Normal distribution [B1]

**(ii)** Discrete uniform distribution [B1]

*e* In part (b)(ii), both the key words 'discrete' and 'uniform' are needed to score the mark.

■ ■ ■

## Question 2

**A statistical model is sometimes used to describe a real-world problem.**

**Explain the process involved in creating such a model.** (4 marks)

## Answer to Question 2

- Identify the real-world problem.
- Devise a statistical model.
- Make predictions about the real-world situations.
- Collect data.
- Obtain expected results.
- Compare the observed data with the expected data.
- Refine the model if necessary.

[Any 4 from this list; B1, B1, B1, B1]

*e* It is worth learning this list. Grade-C candidates often miss out questions that require comments, and as a result they give away marks.

■ ■ ■

## Question 3

**On Christmas Eve, Quickfly Airlines recorded the number of minutes delay in take-off for its flights from a large airport. The results are summarised in the following stem and leaf diagram.**

| Minutes | | | | | | | | | | Totals |
|---|---|---|---|---|---|---|---|---|---|---|
| 0 | 3 | 4 | 4 | 6 | 8 | 9 | | | | (6) |
| 1 | 0 | 1 | 2 | 3 | 5 | 5 | 5 | 8 | 9 | (9) |
| 2 | 1 | 4 | 5 | 5 | 6 | 8 | 9 | | | (7) |
| 3 | 0 | 1 | 7 | 8 | | | | | | (4) |
| 4 | 2 | 4 | 7 | | | | | | | (3) |
| 5 | 3 | 8 | | | | | | | | (2) |

**2 | 3 means 23**

**(a) Find the three quartiles of these data.** (3 marks)

On the same day, the shortest delay on any of Rapidjet's flights from the same airport was 5 minutes. The longest delay was 66 minutes. The three quartiles for this airline were 14, 21 and 57 minutes respectively.

**(b) Draw box plots to represent the data for both airlines.** (6 marks)

**(c) Compare and contrast these two box plots.** (3 marks)

## Answer to Question 3

**(a)** $Q_1 = 11$, $Q_2 = 21$, $Q_3 = 31$ [B1, B1, B1]

**(b)**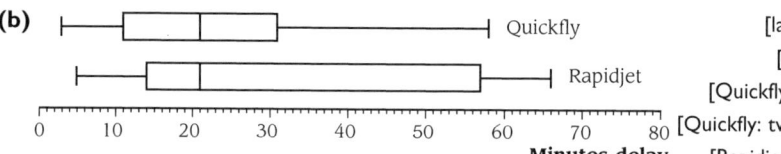

[labels and scale, B1]
[proper shape, M1]
[Quickfly: $Q_1$, $Q_2$, $Q_3$, A1]
[Quickfly: two end values, A1]
[Rapidjet: $Q_1$, $Q_2$, $Q_3$, A1]
[Rapidjet: two end values, A1]

Ensure that the same scale is used for both graphs. This allows comparisons to be made. A sensible scale is usually one that goes up in multiples of any factor of 20, e.g. 1, 2, 4, 5, 10, 20. *Do not* use any scale that is a multiple of 3, for example. It is impossible to read such scales accurately.

If your quartiles were incorrect in part (a), you would be entitled to 'follow through' marks on the box plot.

Labels include 'minutes delay' and the names of the two airlines. The labels may be missing from a grade-C candidate's response.

**(c)** Equal medians, or $Q_3$ bigger for Rapidjet than Quickfly. [B1]

IQR bigger for Rapidjet, or range bigger for Rapidjet. [B1]

Rapidjet is positively skewed ($Q_2 - Q_1 < Q_3 - Q_2$), whereas Quickfly is symmetric ($Q_2 - Q_1 = Q_3 - Q_2$) [B1]

One comment about location, one for spread and one for skewness are the most obvious ways to gain marks here.

## Question 4

In a particular school the distances travelled to school by the teachers were recorded. The shortest distance was 3 miles. The two teachers who travelled the furthest had journeys of 50 and 55 miles. The three quartiles were 17, 23 and 31 miles respectively.

Outliers are values that lie outside the limits:

$$Q_1 - 1.5(Q_3 - Q_1) \text{ and } Q_3 + 1.5(Q_3 - Q_1)$$

(a) Draw a box plot to represent this information. (8 marks)

(b) Comment on the skewness of the distribution. (2 marks)

## Answer to Question 4

(a) $Q_1 - 1.5(Q_3 - Q_1) = 17 - 1.5(31 - 17) = 17 - 21 = -4$ [M1, A1]

$Q_3 + 1.5(Q_3 - Q_1) = 31 + 1.5(31 - 17) = 31 + 21 = 52$ [A1]

So 55 is an outlier. [A1]

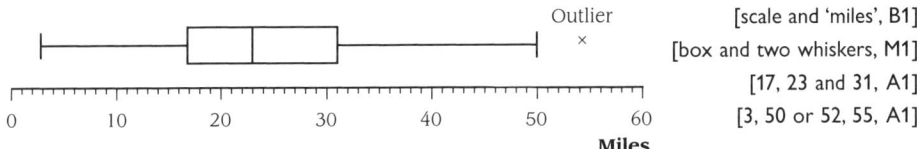

[scale and 'miles', B1]

[box and two whiskers, M1]

[17, 23 and 31, A1]

[3, 50 or 52, 55, A1]

📝 Grade-C candidates sometimes draw the box plot correctly but without showing any of their working for the outliers. This means they are likely to lose all the marks for the calculation. In addition, if a mistake is made in the calculation and it has not been written down, there is nothing for the examiner to follow through, so marks would also be lost on the graph.

(b) $Q_2 - Q_1 = 6$ and $Q_3 - Q_2 = 8$ [M1]

The skew is positive. [A1]

📝 Grade-C candidates are likely to get the method mark for trying to find $Q_2 - Q_1$ and $Q_3 - Q_2$, but they often get the skewness the wrong way round.

■ ■ ■

## Question 5

In a particular week a travel agent kept a record of the length, to the nearest minute, of each of the telephone enquiries he received. The data are summarised in the table below.

| Time/minutes | 2–5 | 6–7 | 8 | 9–12 | 13–17 |
|---|---|---|---|---|---|
| Number of calls | 16 | 20 | 7 | 14 | 11 |

(a) Give a reason to support the use of a histogram to represent these data. (1 mark)

(b) Draw a histogram to illustrate these data. (5 marks)

## Answer to Question 5

**(a)** They are continuous data.                                                                 [B1]
**(b)** Calculating frequency densities:

| Time/minutes | 2–5 | 6–7 | 8 | 9–12 | 13–17 |
|---|---|---|---|---|---|
| **Number of calls** | 16 | 20 | 7 | 14 | 11 |
| **Class widths** | 4 | 2 | 1 | 4 | 5 |
| **Frequency densities** | $\frac{16}{4} = 4$ | $\frac{20}{2} = 10$ | $\frac{7}{1} = 7$ | $\frac{14}{4} = 3.5$ | $\frac{11}{5} = 2.2$ |

[M1, A1]

[scales and labels, B1]
[histogram, with no gaps, M1]
[column heights, A1]

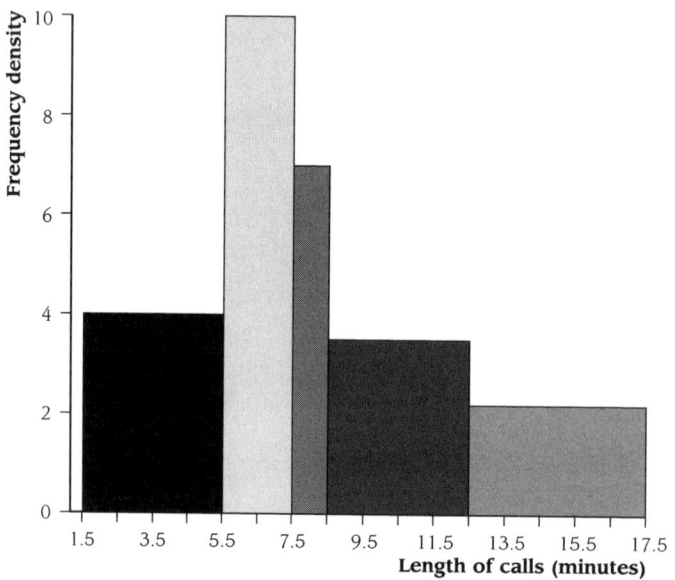

e  Incorrect class widths (typically 3, 2, 1, 3, 4) are the most common reason for marks lost by grade-C candidates.

As with all graphs, a sensible scale is required, not multiples of 3, for example.

If frequencies are plotted, instead of frequency densities, the resulting graph is a bar chart and would score zero. If there are gaps between the columns it means that it is not a histogram, so again no marks would be awarded.

■ ■ ■

## Question 6

**Fifty-five people took part in a school's fun run.**
**The histogram below shows the age distribution of the 55 competitors.**

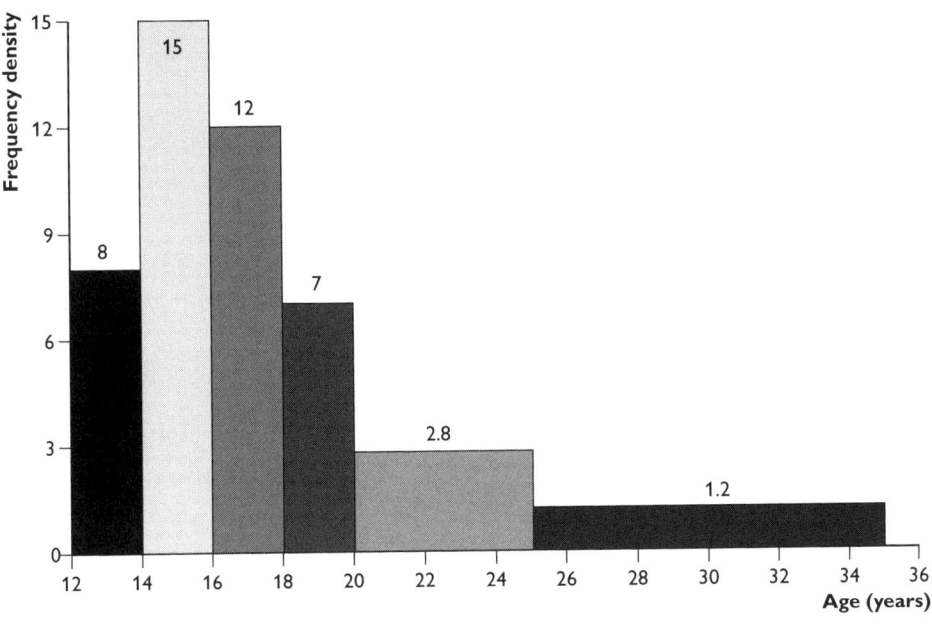

**How many competitors were aged between 20 and 24?** (4 marks)

## Answer to Question 6

Total area of histogram:

= $(2 \times 8) + (2 \times 15) + (2 \times 12) + (2 \times 7) + (5 \times 2.8) + (10 \times 1.2) = 110$ [M1, A1]

But total frequency = 55 (given in the question).

So total area is twice as big as total frequency.

So halving all the separate areas will give the separate frequencies.

Area of column for ages 20 to 24 = $(5 \times 2.8) = 14$ [M1, A1]

Hence there are $\dfrac{14}{2}$ = 7 competitors

*e* Here the histogram is based on *proportion*, rather than the simplest case of using frequency to *equal* area.

■ ■ ■

## Question 7

**The number of gas barbecues sold per day at a garden centre was recorded over a period of 10 consecutive days. The results are shown below.**

**10, 5, 7, 12, 27, 31, 9, 3, 8, 13**

**(a) Find the median and interquartile range of these data.** (3 marks)

**(b) Calculate the mean and standard deviation of the data.** (5 marks)

## Answer to Question 7

**(a)** Rearranging the data in order: 3, 5, 7, 8, 9, 10, 12, 13, 27, 31

Median, $Q_2$ = 9.5 (halfway between 5th and 6th values) [B1]

$Q_1$ = 7, $Q_3$ = 13 [B1 for both values]

IQR = 13 – 7 = 6 [B1]

**(b)** $\sum x = 3 + 5 + 7 + 8 + 9 + 10 + 12 + 13 + 27 + 31 = 125$ [M1]

$\sum x^2 = 3^2 + 5^2 + 7^2 + \ldots + 27^2 + 31^2 = 2331$ [M1]

mean $= \dfrac{\sum x}{n} = \dfrac{125}{10} = 12.5$ [A1]

standard deviation $= \sqrt{\dfrac{\sum x^2}{n} - \left(\dfrac{\sum x}{n}\right)^2}$

$= \sqrt{\dfrac{2331}{10} - \left(\dfrac{125}{10}\right)^2}$

$= \sqrt{233.1 - 12.5^2}$

$= \sqrt{76.85}$ [A1, for 76.85]

$= 8.77$ [A1]

 A grade-C candidate might stop at the value 76.85, forgetting to take the square root. Another common mistake is to use $125^2$ as the value for $\sum x^2$. This would lose all 3 marks for calculating the standard deviation.

■ ■ ■

## Question 8

The values of the daily sales, to the nearest £, taken at the toy department of a large store are summarised in the table below.

| Sales/£ | Number of days |
|---------|----------------|
| 1–500 | 9 |
| 501–800 | 23 |
| 801–1100 | 48 |
| 1101–1300 | 91 |
| 1301–1500 | 139 |

**(a)** Use interpolation to estimate the median and interquartile range of daily sales.

(5 marks)

**(b) Estimate the mean and standard deviation of these data.** (6 marks)

**(c) One coefficient of skewness is given by:**

$$\frac{3(\text{mean} - \text{median})}{\text{standard deviation}}$$

Evaluate this coefficient for the above data. (2 marks)

**(d) The manager of the toy department wants to compare last year's sales with other years. State whether the manager should use the median and the interquartile range *or* the mean and the standard deviation to compare daily sales. Give a reason for your answer.** (2 marks)

## Answer to Question 8

| Sales/£,x | Number of days, $f$ | Cumulative frequency | Mid-interval values, $x$ | $fx$ | $fx^2$ |
|---|---|---|---|---|---|
| 1–500 | 9 | 9 | 250.5 | 2254.5 | 564752.25 |
| 501–800 | 23 | 32 | 650.5 | 14961.5 | 9732455.75 |
| 801–1100 | 48 | 80 | 950.5 | 45624.0 | 43365612 |
| 1101–1300 | 91 | 171 | 1200.5 | 109245.5 | 131149222.8 |
| 1301–1500 | 139 | 310 | 1400.5 | 194669.5 | 272634634.8 |
| | $\sum f = 310$ | | | $\sum fx = 366755$ | $\sum fx^2 =$ 457446677.5 |

**(a)** Median = 155th value

From the cumulative frequencies, this lies in the interval 1101–1300.

median: $\dfrac{(m - 1100.5)}{(1300.5 - 1100.5)} = \dfrac{(155 - 80)}{(171 - 80)}$       [M1]

$\dfrac{m - 1100.5}{200} = \dfrac{(75)}{(91)}$

median, $m = 1100.5 + \dfrac{200 \times 75}{91} = £1265.34$       [A1]

Similarly, $Q_1 = 800.5 + \dfrac{300 \times 45.5}{48} = £1084.88$       [A1]

and $Q_3 = 1300.5 + \dfrac{200 \times 61.5}{139} = £1388.99$       [A1]

Interquartile range = $Q_3 - Q_1 = 1388.99 - 1084.88 = £304.11$       [B1]

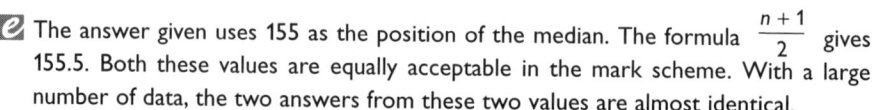 The answer given uses 155 as the position of the median. The formula $\dfrac{n+1}{2}$ gives 155.5. Both these values are equally acceptable in the mark scheme. With a large number of data, the two answers from these two values are almost identical.

**(b)** From the table above: mean, $\bar{x} = \dfrac{\sum fx}{\sum f} = \dfrac{366\,755}{310} = £1183.08$   [M1 for $\sum fx$]

                          [M1, A1]

standard deviation $= \sqrt{\dfrac{\sum fx^2}{\sum f} - \left(\dfrac{\sum fx}{\sum f}\right)^2} = \sqrt{\dfrac{457\,446\,677.6}{310} - \left(\dfrac{366\,755}{310}\right)^2}$   [M1 for $\sum fx^2$]

                              [M1]

$\qquad\qquad\qquad = \sqrt{(1\,475\,634.444 - 1\,399\,680)} = \sqrt{(75\,954.63)} = £275.60$   [A1]

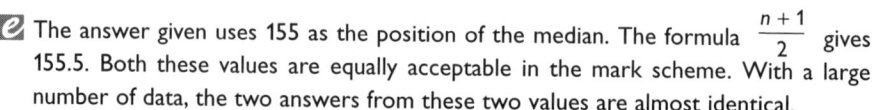 C-grade candidates often lose accuracy marks by rounding off values too early in the calculation. For example, if you round off the mean to £1183 and then use this value in the formula for standard deviation, you get an answer of £275.94, which is not accurate.

Note that this particular question involves a lot of basic work on the calculator to find the values of $\sum fx$ and $\sum fx^2$. In most cases you will be given these values as part of the information in the question.

**(c)** $\dfrac{3(\text{mean} - \text{median})}{\text{standard deviation}} = \dfrac{3(1183.08 - 1265.34)}{275.60} = -0.895$   [M1, A1]

**(d)** Use the median and the IQR because the data are skewed.   [B1, B1]

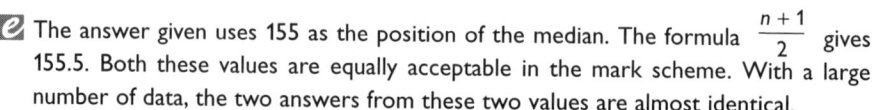 Make sure that *both* terms (median and IQR) are named.

■ ▨ ■

# Question 9

**At a clinic, the weights of 50 small children were taken. The results are summarised in the table below.**

| Weight/kg | Frequency |
|---|---|
| 15.6–16.0 | 4 |
| 16.0–16.5 | 7 |
| 16.5–19.0 | 0 |
| 19.0–19.6 | 10 |
| 19.6–20.0 | 18 |
| 20.0–21.0 | 11 |

**Use the coding y = 10(x – 15) to find an estimate for the mean and standard deviation of the weights of the children.**

**(You may use $\sum fy^2 = 94\,586.75$)**       (6 marks)

## Answer to Question 9

| Weight/kg | Frequency, $f$ | Mid-point, $x$ | $y = 10(x - 15)$ | $fy$ |
|-----------|-----------|-----------|-----------|-----------|
| 15.6–16.0 | 4 | 15.8 | 8 | 32 |
| 16.0–16.5 | 7 | 16.25 | 12.5 | 87.5 |
| 16.5–19.0 | 0 | 17.75 | 27.5 | 0 |
| 19.0–19.6 | 10 | 19.3 | 43 | 430 |
| 19.6–20.0 | 18 | 19.8 | 48 | 864 |
| 20.0–21.0 | 11 | 20.5 | 55 | 605 |
| | $\sum f = 50$ | | | $\sum fy = 2018.5$ |

$$\text{mean of } y = \frac{\sum fy}{\sum f} = \frac{2018.5}{50} = 40.37 \qquad \text{[M1]}$$

$$\text{standard deviation of } y = \sqrt{\frac{\sum fy^2}{\sum f} - \left(\frac{\sum fy}{\sum f}\right)^2} = \sqrt{\frac{94586.75}{50} - \left(\frac{2018.5}{50}\right)^2} \qquad \text{[M1, A1]}$$

$$= \sqrt{(1891.735 - 40.37^2)} = \sqrt{261.9981} = 16.1864$$

Decoding: if $y = 10(x - 15)$, then $x = \dfrac{y}{10} + 15$

$$\text{mean of } x = \frac{\text{mean of } y}{10} + 15 = 4.037 + 15 = 19.04 \text{ kg (3 s. f.)} \qquad \text{[M1, A1]}$$

$$\text{standard deviation of } x = \frac{\text{standard deviation of } y}{10} = 1.62 \text{ kg} \qquad \text{[A1]}$$

✐ Coding questions are quite rare. Grade-A candidates usually remember how the decoding works. Some grade-C candidates waste time by recalculating $\sum fy^2$ even though it is given in the question.

■ ■ ■

## Question 10

A fair dice has six faces, numbered 1, 2, 2, 3, 4, 4. The dice is rolled twice and the number showing on the uppermost face is recorded each time.
Find the probability that the sum of the two numbers recorded is at least 6. (5 marks)

## Answer to Question 10

[M1, M1, A1]

|  |  |  |  |  |  |  |  |
|---|---|---|---|---|---|---|---|
| 4 | | 5 | 6 | 6 | 7 | 8 | 8 |
| 4 | | 5 | 6 | 6 | 7 | 8 | 8 |
| **Score on** 3 | | 4 | 5 | 5 | 6 | 7 | 7 |
| **2nd dice** 2 | | 3 | 4 | 4 | 5 | 6 | 6 |
| 2 | | 3 | 4 | 4 | 5 | 6 | 6 |
| 1 | | 2 | 3 | 3 | 4 | 5 | 5 |
| | | 1 | 2 | 2 | 3 | 4 | 4 |
| | | | | **Score on 1st dice** | | | |

$$\text{probability (total is 6 or more)} = \frac{17}{36} \qquad \text{[M1, A1]}$$

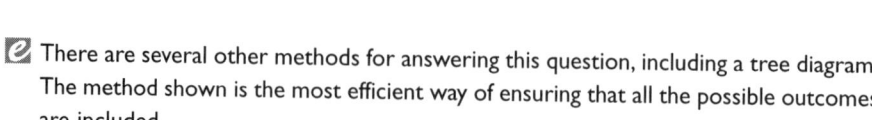 There are several other methods for answering this question, including a tree diagram. The method shown is the most efficient way of ensuring that all the possible outcomes are included.

Grade-C candidates who try other methods often either miss one or more of the possible outcomes or count one or more of them twice. This is particularly true if they try to use a method that involves listing all the ways of getting a total of six or more, perhaps forgetting that (2,4) and (4,2) both have to be included and that there are several ways to get those particular pairs because of the repeated numbers on the dice.

■ ■ ■

# Question 11

The events A and B are such that $P(\mathbf{A}) = \dfrac{1}{4}$, $P(\mathbf{B}) = \dfrac{2}{5}$ and $P(\mathbf{A} \cap \mathbf{B}) = \dfrac{1}{10}$.

(a) Represent these probabilities in a Venn diagram. (4 marks)

Hence, or otherwise, find:

(b) $P(\mathbf{A} \cup \mathbf{B})$ (1 mark)

(c) $P(\mathbf{A} \mid \mathbf{B}')$ (2 marks)

## Answer to Question 11

(a)

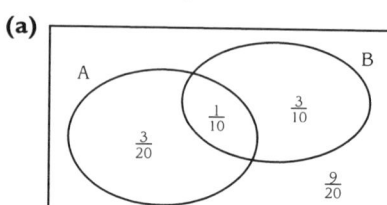

[2 intersecting loops, M1]

[subtracting $\dfrac{1}{10}$, M1]

[$\dfrac{3}{20}$ and $\dfrac{3}{10}$, A1]

[box drawn and $\dfrac{9}{20}$, B1]

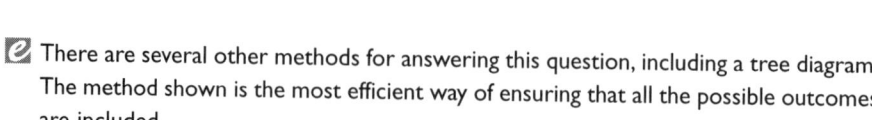 Grade-C candidates often forget to include the value 9/20 outside the two loops.

Another common error is failure to subtract the value of $P(A \cap B)$ from both $P(A)$ and $P(B)$ and just putting in the values of $P(A)$ and $P(B)$ in the two outer parts of the loops.

(b) $P(A \cup B) = \dfrac{3}{20} + \dfrac{1}{10} + \dfrac{3}{10} = \dfrac{11}{20}$ [B1]

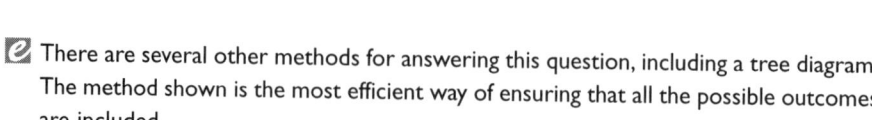 This value would have been calculated already in order to work out the outer value of 9/20.

(c) $P(A \mid B') = \dfrac{P(A \cap B')}{P(B')} = \dfrac{{}^{3}/_{20}}{{}^{3}/_{5}} = \dfrac{{}^{3}/_{20}}{{}^{12}/_{20}} = \dfrac{1}{4}$ [M1, A1]

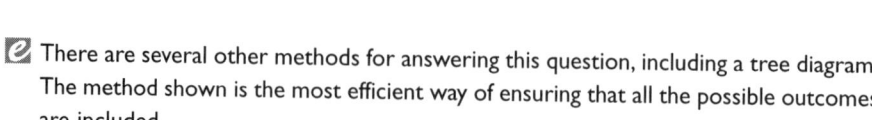 The most likely error here is to divide by $P(A')$ instead of $P(B')$.

■ ■ ■

## Question 12

In a survey, 50 adults were asked which of the holiday destinations Spain, Turkey and Florida they had visited. The results showed that 29 had visited Spain, 15 had visited Turkey, 16 had visited Florida, 9 had visited both Spain and Turkey, 8 had visited both Spain and Florida, 5 had visited both Turkey and Florida and 2 had visited all three destinations.

(a) Draw a Venn diagram to represent these data. (6 marks)

(b) One of the adults was then selected at random. Find the probability that he/she had visited:
   (i)   at least one of the destinations (2 marks)
   (ii)  only Spain (1 mark)
   (iii) only one of the destinations (2 marks)
   (iv) Spain, given that he/she had visited only one of the destinations (2 marks)

## Answer to Question 12

(a)

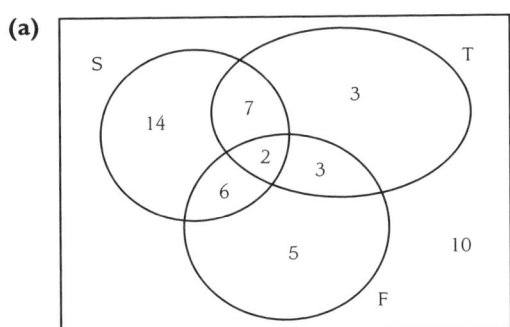

[2, B1]
[7, 6, 3, M1, A1]
[14, 3, 5, M1, A1]
[Box and 10, B1]

*e* This is a common type of question. The key is to start with the one fact in the question that relates to a single area on the Venn diagram. Here it is the number of adults that had visited all three destinations. Then work your way outwards, taking note that all the other facts relate to more than one area. For example, '29 had visited Spain' tells you that it is the total of all four areas inside the Spain loop that must add up to 29. Hence a number of subtractions are needed to complete the diagram. The most common mistake is to put 29 into the part of the Spain loop that has the correct value of 14 in it. Candidates often claim that this kind of question is ambiguous in its wording. It is not. Don't forget to draw the box and put the 10 in the diagram.

(b) (i) $P$(at least one destination) $= \dfrac{14 + 7 + 3 + 6 + 2 + 3 + 5}{50} = \dfrac{40}{50} = \dfrac{4}{5}$

or $1 - \dfrac{10}{50} = \dfrac{4}{5}$       [M1, A1]

*e* Grade-C candidates may lose marks in part (a) by failing to carry out the necessary subtractions and hence have values such as 29, 15 and 16 etc. in the diagram. This error would be penalised in part (a), but it is still possible to gain all the marks in part (b), as long as subsequent answers are consistent with the Venn diagram that you have drawn.

**(ii)** $P(\text{only Spain}) = \dfrac{14}{50} = \dfrac{7}{25}$ [B1]

**(iii)** $P(\text{only one of the destinations}) = \dfrac{14+3+5}{50} = \dfrac{22}{50} = \dfrac{11}{25}$ [M1, A1]

*e* Parts (b)(ii) and (iii) are often well answered.

**(iv)** $P(\text{Spain, given visited only one}) = P(S \mid \text{only 1}) = \dfrac{P(S \cap \text{only one})}{P(\text{only one})}$ [M1]

$\dfrac{\text{answer to part (b)(ii)}}{\text{answer to part (b)(iii)}} = \dfrac{\frac{7}{25}}{\frac{11}{25}} = \dfrac{7}{11}$ [A1]

*e* Grade-C candidates may not realise that this is a conditional probability. The key words in the question are 'given that'. There are some occasions when these words do not specifically appear, but the context makes it clear that there is a conditional probability.

One common error occurs with candidates who prefer to write the answer in decimal form:

$\dfrac{7}{11} = 0.6363636$

The answer 0.64 would lose the accuracy mark. It is much better to leave the exact fractional answer.

■ ■ ■

# Question 13

The events **X** and **Y** are such that $P(\mathbf{X}) = \dfrac{3}{5}$, $P(\mathbf{Y}) = \dfrac{1}{3}$ and $P(\mathbf{X}|\mathbf{Y}') = \dfrac{9}{10}$.

**(a) Find:**
   **(i)** $P(\mathbf{X} \cap \mathbf{Y}')$          (2 marks)
   **(ii)** $P(\mathbf{X} \cap \mathbf{Y})$         (2 marks)
   **(iii)** $P(\mathbf{X} \cup \mathbf{Y})$        (2 marks)
   **(iv)** $P(\mathbf{X}|\mathbf{Y})$          (1 mark)

**(b) State, with a reason, whether or not X and Y are:**
   **(i)** mutually exclusive     (2 marks)
   **(ii)** independent         (2 marks)

# Answer to Question 13

**(a) (i)** $P(X|Y') = \dfrac{P(X \cap Y')}{P(Y')}$ [M1]

So, $P(X \cap Y') = P(Y') \times P(X|Y')$

$= \dfrac{2}{3} \times \dfrac{9}{10} = \dfrac{18}{30} = \dfrac{3}{5}$ [A1]

🖉 A common mistake is to calculate $P(X) \times P(Y')$.

This would be correct only if $X$ and $Y$ were known to be independent. At this stage in the question there is nothing to indicate that they are independent, so we cannot use that method.

**(ii)** Either:

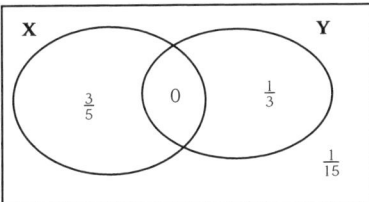

So, $P(X \cap Y) = 0$

Or:

$$P(X \cap Y) = P(X) - P(X \cap Y')$$ [M1]

$$= \frac{3}{5} - \frac{3}{5} = 0$$ [A1]

🖉 C-grade candidates may not realise that the answer to part (i) is the left-hand area in loop X. If this is put into a Venn diagram, then the rest of the question is simple.

**(iii)** $P(X \cup Y) = \frac{3}{5} + 0 + \frac{1}{3} = \frac{14}{15}$ [M1, A1]

**(iv)** $P(X|Y) = \dfrac{P(X \cap Y)}{P(Y)} = 0$ [B1]

🖉 The last two parts are straightforward, particularly if a Venn diagram has been drawn.

**(b) (i)** $P(X \cap Y) = 0$, so X and Y are mutually exclusive. [M1, A1]

**(ii)** If X and Y are independent, then $P(X) \times P(Y) = P(X \cap Y)$. [M1]

$P(X) = \frac{3}{5}$ and $P(Y) = \frac{1}{3}$, so $P(X) \times P(Y) = \frac{3}{5} \times \frac{1}{3} = \frac{1}{5}$

But $P(X \cap Y) = 0$

X and Y are not independent because:

$$P(X) \times P(Y) \neq P(X \cap Y)$$ [A1]

🖉 In this part, grade-C candidates would probably achieve the method mark for attempting to show that, for independence:

$$P(X) \times P(Y) = P(X \cap Y)$$

Often the A1 mark is lost because the candidate fails to state the required condition.

In both sections of part (b), if a candidate has made the mistake suggested in part (a)(i), then these two conclusions are bound to be wrong.

## Question 14

A child has a bag containing **11** red sweets and **5** green sweets. The child randomly chooses a sweet, looks at the colour and then eats the sweet. A second sweet is then chosen at random, its colour noted and then the child eats that sweet.

**(a)** Draw a tree diagram to represent this information. (3 marks)

**(b)** Find the probability that:

    **(i)** the second sweet is green (2 marks)

    **(ii)** both sweets are green, given that the second sweet is green (2 marks)

## Answer to Question 14

**(a)**

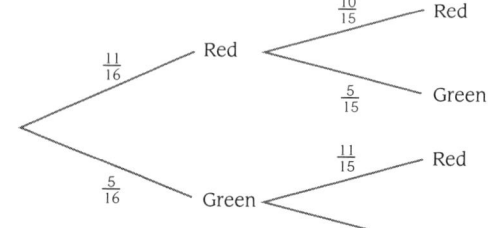

1st sweet      2nd sweet      [tree with correct number of branches, M1]

$[\frac{11}{16}$ and $\frac{5}{16}$, A1]

[all correct, with labels, A1]

📝 This part of the question is usually answered well, as long as the second set of branches have the denominators reduced by one.

**(b)** $P(\text{second is green}) = P(\text{red}) \times P(\text{green}) + P(\text{green}) \times P(\text{green})$

$$= \frac{11}{16} \times \frac{5}{15} + \frac{5}{16} \times \frac{4}{15} = \frac{11}{48} + \frac{4}{48} = \frac{15}{48} = \frac{5}{16} \qquad \text{[M1, A1]}$$

**(c)** $P(\text{both green} \mid \text{2nd is green}) = \dfrac{\frac{5}{16} \times \frac{4}{15}}{\frac{5}{16}} = \dfrac{4}{15}$      [M1, A1]

📝 In part (c), a C-grade candidate is likely to get the numerator correct but may use 4/15 (from the tree) as the denominator.

■ ■ ■

## Question 15

A statistics student drew the following scatter diagram.

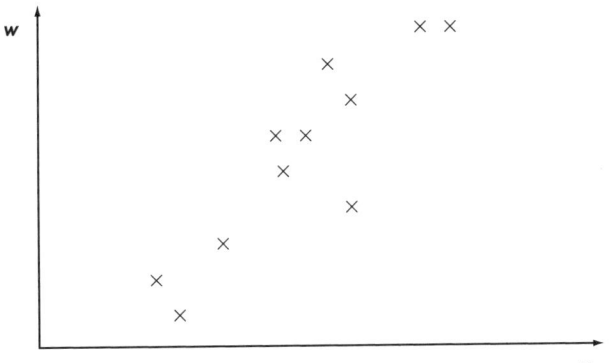

State, with a reason, which of the following product–moment correlation coefficients corresponds to the diagram:

0.16    −0.72    0.69                                                                    (2 marks)

## Answer to Question 15

0.69                                                                                          [B1]

Reason: (one of the following)
- as $v$ increases, $w$ also increases
- most points lie in the 1st and 3rd quadrants                                             [B1]

📝 Grade-C candidates may lose the mark for the reason, by giving one of the following reasons.
- It shows positive correlation. (Here the value of 0.69 is an indicator of positive correlation. The reason given needs to explain *why* we know the correlation is positive.)
- As $x$ increases, $y$ also increases. (The variables are $v$ and $w$, not $x$ and $y$.)
- As one increases so does the other. (This is not specific enough. You must name the correct variables, $v$ and $w$.)

## Question 16

Members at a local gym were invited to take part in the monthly fitness challenge, consisting of counting the number of squats, $x$, that a member could do in 1 minute and the number of press-ups, $y$, that could be done in 1 minute. Twelve members took part in the challenge. The results are summarised below.

$$\sum x = 456, \sum x^2 = 17\,992, \sum xy = 19\,269, \sum y = 496, \sum y^2 = 20\,936$$

(a) Find the exact values of $S_{xx}$, $S_{yy}$ and $S_{xy}$.                              (4 marks)
(b) Calculate, to 3 decimal places, the product–moment correlation coefficient
between $x$ and $y$.                                                                        (2 marks)
(c) Give an interpretation of your coefficient.                                            (2 marks)

## Answer to Question 16

**(a)** $S_{xx} = \sum x^2 - \dfrac{\left(\sum x\right)^2}{n} = 17\,992 - \dfrac{456^2}{12} = 664$     [M1, A1]

$S_{yy} = \sum y^2 - \dfrac{\left(\sum y\right)^2}{n} = 20\,936 - \dfrac{496}{12} = 434\dfrac{2}{3}$     [A1]

$S_{xy} = \sum xy - \dfrac{\left(\sum x \sum y\right)}{n} = 19\,269 - \dfrac{456 \times 496}{12} = 421$     [A1]

**e** The most likely reason for C-grade candidates losing a mark is writing the value of $S_{yy}$ as 434.67 instead of giving it as an exact fraction. If the question asks for *exact values*, rounded-off decimals will lose a mark.

*All* the required formulae for the calculations are stated in the examination formula book. Full marks can be gained simply through careful substitution and accurate calculator work.

**(b)** PMCC $= \dfrac{S_{xy}}{\sqrt{\left(S_{xx}S_{yy}\right)}} = \dfrac{421}{\sqrt{\left(664 \times 434\dfrac{2}{3}\right)}} = 0.784$     [M1, A1]

**e** Failing to follow the instruction about 3 decimal places is the most likely way to lose a mark. Again, the formula for the product–moment correlation coefficient is given in the formula book.

**(c)** 0.784 suggests a strong positive correlation.     [B1]

This suggests that those who can do the most squats can also do the most press-ups.     [B1]

**e** The grade-C candidate might miss the second mark. All attempts to interpret values should have a comment that is written *in the context* of the variables mentioned in the question.

■ ■ ■

## Question 17

**A golf coach monitored the progress of a pupil over a period of 10 weeks. At the end of each week the pupil played a round of golf (18 holes) and the number of strokes, y, the pupil took to complete the round was recorded. The coach also recorded the number of hours, x, the pupil practised each week. The data are shown in the table below.**

| x | 12 | 18 | 7 | 13 | 17 | 15 | 10 | 6 | 23 | 13 |
|---|----|----|---|----|----|----|----|---|----|----|
| y | 84 | 75 | 89 | 86 | 74 | 82 | 90 | 91 | 70 | 81 |

**(a) Plot these data on a scatter diagram.**     (3 marks)

**(b) Find the equation of the regression line of y on x in the form y = a + bx.**

(You may use $\sum x^2 = 2034$, $\sum xy = 10696$) (9 marks)

(c) Give an interpretation of the value of $a$ and the value of $b$ from your regression
equation. (2 marks)

(d) Predict the score that the pupil might achieve with:
(i) 20 hours of practice
(ii) 30 hours of practice
In each case, comment on the validity of your estimate. (4 marks)

## Answer to Question 17

(a)

[sensible scales and labels, B1]

[all points correctly plotted, B2]

(allow B1 if 8 or 9 are correct)

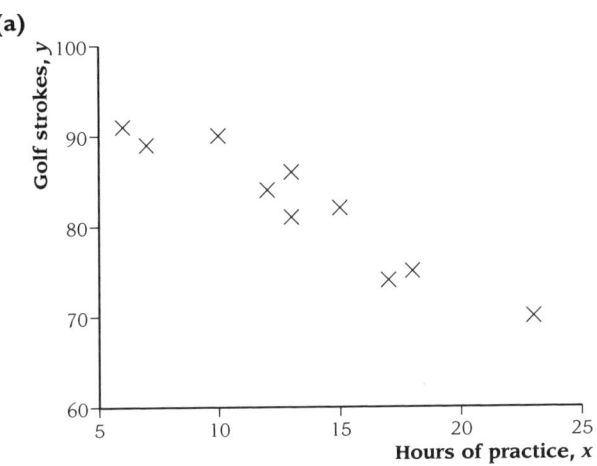

There are two aspects to 'sensible scale' here. First, the scales should not be multiples of 3. Second, the y-axis should not start at 0, because this would leave a lot of redundant space at the bottom of the graph. In this question, starting the axis at 0 may not be penalised but in some questions (e.g. if the values of the variable are between 150 and 200), a mark would be lost for starting the axis at 0. It may even result in the loss of a second mark because it might be difficult to check the accuracy of the points if they are all bunched close together on the graph.

(b) $\sum x = 134$, $\sum y = 822$ [B1, B1]

$$S_{xy} = \sum xy - \frac{\left(\sum x \sum y\right)}{n} = 10696 - \frac{134 \times 822}{10} = -318.8$$ [M1, A1]

$$S_{xx} = \sum x^2 - \frac{\left(\sum x\right)^2}{n} = 2034 - \frac{134^2}{10} = 238.4$$ [A1]

$$b = \frac{S_{xy}}{S_{xx}} = \frac{-318.8}{238.4} = -1.33725$$ [M1, A1]

$$a = \bar{y} - b\bar{x} = \frac{822}{10} - \left(-1.33725 \times \frac{134}{10}\right) = 82.2 + 17.91915 = 100.12$$ [A1]

The equation is $y = 100 - 1.34x$ (*a* and *b* correct to 3 significant figures) [B1]

 Unless the question says otherwise, the rule of thumb is to give the values of *a* and *b* correct to 3 significant figures. It is common for candidates to lose a mark if they fail to do this.

All the formulae used in this answer are found in the examination formula book.

Grade-C candidates often round off too early and might give the value of *b* as −1.34. This would score all the marks for *b*, but it often leads to an inaccurate value for *a*. In this example, the value of *a* would be **100.16** instead of **100.12**. Such a small difference may be ignored, but often the difference is much more significant and would be penalised.

**(c)** Interpretations:
- 100 is the estimated score if the pupil practised for 0 hours. [B1]
- 1.34 is the amount that the score decreases by for each extra hour of practice. [B1]

 Grade-C candidates often simply state that *a* is the intercept on the y-axis and *b* is the gradient. Any interpretation must be in the context of the question, using the named variables.

**(d)** If $x = 20$ then $y = 100 - 1.34 \times 20 = 73.2$

So the predicted score is 74. [B1]

This is reliable because 20 is within the range of the collected data. [B1]

If $x = 30$, $y = 100 - 1.34 \times 30 = 59.8$

So the predicted score is 60. [B1]

This is unreliable because it is outside the range. The linear model cannot continue indefinitely. If it did, then, if $x = 60$ for example, the pupil would complete the round of golf in 20 shots, which would need to include 16 holes in 1! [B1]

 Rounds of golf can only have scores that are whole numbers, hence the rounding off. The reliability is determined by whether the *x* values are within the range of the original data.

■ ■ ■

# Question 18

A project was set up to investigate the relationship between the grams of fertiliser, *x*, given to eight tomato plants and the yield, *y*, in kilograms, for each plant. The data below summarise the results of the investigation.

The data are coded so that $s = x - 2$ and $t = y - 5$.

$$\sum s = 14.7, \sum t = 13.7, \sum st = 39.92, \sum s^2 = 53.69, \sum t^2 = 32.77$$

(a) Calculate the values of $S_{ss}$, $S_{tt}$ and $S_{st}$, giving your answers correct to 3 decimal
   places. (4 marks)
(b) Find the equation of the regression line of $t$ on $s$ in the form $t = a + bs$. (5 marks)
(c) Hence find the equation of the regression line of $y$ on $x$. (3 marks)
(d) Calculate the product–moment correlation coefficient between $s$ and $t$. (3 marks)
(e) Hence find the product–moment correlation coefficient between $x$ and $y$. (1 mark)

## Answer to Question 18

(a) $S_{ss} = 53.69 - \dfrac{14.7^2}{8} = 26.67875 = 26.679$ [M1, A1]

$S_{tt} = 32.77 - \dfrac{13.7^2}{8} = 9.30875 = 9.309$ [A1]

$S_{st} = 39.92 - \dfrac{14.7 \times 13.7}{8} = 14.74625 = 14.746$ [A1]

💡 All these formulae are in the examination formula book. The most likely error for
C-grade candidates is to ignore the instruction concerning 3 decimal places.

(b) $b = \dfrac{S_{st}}{S_{ss}} = \dfrac{14.74625}{26.67875} = 0.552734$ [M1, A1]

$a = \bar{t} - b\bar{s} = \dfrac{13.7}{8} - 0.552734 \times \dfrac{14.7}{8} = 0.69685$ [M1, A1]

equation: $t = 0.697 + 0.553s$ [B1]

💡 Grade-C candidates might round the $b$ value to 0.553 before calculating the value of
$a$. This leads to $a = 0.696$ instead of 0.697, which would lose a mark. To get the final
mark the equation must contain $s$ and $t$, and not $x$ and $y$.

(c) $t = 0.697 + 0.553s$

So substituting the two codes gives:

$(y - 5) = 0.697 + 0.553(x - 2)$ [M1]

$y - 5 = 0.697 + 0.553x - 1.106$ [A1]

$y = 4.591 + 0.553x$ [A1]

💡 A simple substitution is needed to carry out the decoding.

(d) PMCC $= \dfrac{S_{st}}{\sqrt{(S_{ss}S_{tt})}} = \dfrac{14.74625}{\sqrt{(26.67875 \times 9.30875)}}$ [M1, A1]

$= 0.936$ [A1]

💡 Once again, if the values of $S_{ss}$, $S_{tt}$ and $S_{st}$ are rounded off to 3 significant figures earlier
and the rounded-off values used in this part, the answer is 0.932 instead of the
required 0.936. This would not score the last mark.

(e) $r = 0.936$ [B1]

 Coding has no effect on the correlation coefficient, so there is no decoding to do here. The correlation coefficient for $x$ and $y$ is the same as for $s$ and $t$.

■ ■ ■

## Question 19

**A discrete random variable $X$ has the probability function shown in the table below.**

| $x$ | 0 | 1 | 2 | 3 |
|---|---|---|---|---|
| $P(X = x)$ | 0.4 | 0.1 | 0.2 | 0.3 |

Find:
(a) $P(1 \leq X < 3)$          (2 marks)
(b) $F(1.75)$          (1 mark)
(c) $E(X)$          (2 marks)
(d) $E(3X - 4)$          (2 marks)
(e) $Var(X)$          (4 marks)
(f) $Var(3X - 4)$          (2 marks)

## Answers to Question 19

**(a)** $P(1 \leq X < 3) = P(X = 1) + P(X = 2)$          [M1]

$$= 0.1 + 0.2 = 0.3$$          [A1]

 Adding on $P(X = 3)$ is the most likely error that a C-grade candidate might make here.

**(b)** $F(1.75) = P(X \leq 1.75) = P(X = 0) + P(X = 1) = 0.4 + 0.1 = 0.5$          [B1]

 Grade-C candidates sometimes miss out this question because they are not confident about dealing with the decimal value. Occasionally they round off **1.75** to the nearest whole number and find $F(2)$.

**(c)** $E(X) = \sum x P(X = x) = (0 \times 0.4) + (1 \times 0.1) + (2 \times 0.2) + (3 \times 0.3)$          [M1]

$= 0 + 0.1 + 0.4 + 0.9 = 1.4$          [A1]

 This is usually answered well. Candidates should write out the first line of the solution with all the probabilities multiplied by the values of the variable because this scores the method mark. Without this line a wrong answer would score 0.

**(d)** $E(3X - 4) = 3E(X) - 4$          [M1]

$= 3 \times 1.4 - 4 = 0.2$          [A1]

**(e)** $E(X^2) = (0^2 \times 0.4) + (1^2 \times 0.1) + (2^2 \times 0.2) + (3^2 \times 0.3)$          [M1]

$= 0 + 0.1 + 0.8 + 2.7 = 3.6$          [A1]

$\text{Var}(X) = E(X^2) - (E(X))^2 = 3.6 - 1.4^2$ [M1]

$= 3.6 - 1.96 = 1.64$ [A1]

*e* Forgetting to subtract $(E(X))^2$ is the most common mistake, which produces the incorrect answer 3.6.

Another mistake that grade-C candidates sometimes make is to square the wrong value in the calculation of $E(X^2)$. For example, $(0 \times 0.4^2) + (1 \times 0.1^2) + \dots$ is sometimes seen, as is $(0 \times 0.4)^2 + (1 \times 0.1)^2 + \dots$

**(f)** $\text{Var}(3X - 4) = 3^2 \text{Var}(X)$ [M1]

$= 3^2 \times 1.64 = 14.76$ [A1]

*e* Grade-C candidates do not usually make the mistake of finding $3\text{Var}(X) - 4$, although they do sometimes find $3\text{Var}(X)$ instead of $3^2\text{Var}(X)$.

■ ■ ■

## Question 20

**A discrete random variable $X$ has the probability function shown in the table below, where $h$ and $k$ are constants.**

| $x$ | 1 | 2 | 3 | 4 |
|---|---|---|---|---|
| $P(X = x)$ | 0.3 | $h$ | $k$ | 0.1 |

**(a) Given that $E(X) = 2.4$, find the value of $h$ and the value of $k$.** (5 marks)
**(b) Show that $\text{Var}(X) = 1.04$.** (3 marks)
**A discrete random variable $Y$ is such that $E(Y) = 3.6$ and $\text{Var}(Y) = 1.7$. Evaluate:**
**(c) $E(5X - 3Y)$** (2 marks)
**(d) $\text{Var}(5X - 3Y)$** (2 marks)

## Answer to Question 20

**(a)** Equation 1, using $\sum p = 1$: $0.3 + h + k + 0.1 = 1$ [M1]

$h + k = 0.6$ [A1]

Equation 2, using $E(X) = \sum xp$: $(1 \times 0.3) + (2 \times h) + (3 \times k) + (4 \times 0.1) = 2.4$ [M1]

$0.3 + 2h + 3k + 0.4 = 2.4$

$2h + 3k = 1.7$ [A1]

Solving equations 1 and 2 simultaneously:

$2h + 3k = 1.7$

$2h + 2k = 1.2$

$k = 0.5$

$h = 0.1$ [B1 (both values needed)]

*e* Grade-C candidates sometimes spend a lot of time trying to solve this problem using trial and improvement because they do not see that there are two equations that can

be formed from the information in the question. Occasionally they simply assume that $h$ and $k$ must each be 0.3 to make the probabilities add up to 1.

**(b)** $E(X^2) = (1^2 \times 0.3) + (2^2 \times 0.1) + (3^2 \times 0.5) + (4^2 \times 0.1)$         [M1]

$\qquad\qquad = 0.3 + 0.4 + 4.5 + 1.6 = 6.8$         [A1]

$\qquad \text{Var}(X) = 6.8 - 2.4^2 = 6.8 - 5.76 = 1.04$         [A1]

*e* Because this answer is given, a fully correct solution is needed. Examiners can easily spot any attempts to fudge the solution to make the answer appear from incorrect working.

**(c)** $E(5X - 3Y) = 5E(X) - 3E(Y)$         [M1]

$\qquad\qquad = 5 \times 2.4 - 3 \times 3.6 = 12 - 10.8 = 1.2$         [A1]

*e* Such questions are usually done well.

**(d)** $\text{Var}(5X - 3Y) = 5^2\text{Var}(X) + 3^2\text{Var}(Y)$         [M1]

$\qquad\qquad = 25 \times 1.04 + 9 \times 1.7$

$\qquad\qquad = 26 + 15.3 = 41.3$         [A1]

*e* Grade-C candidates usually remember to square the 5 and the 3 but occasionally forget that the two variances must be *added*. The first line in the above solution must be completely correct to gain any marks in this part of the question.

■ ■ ■

## Question 21

The random variable $X$ has the probability function:

$$P(X=x)=\begin{cases} kx & x=1, 2 \\ k(x-1) & x=3, 4, 5 \\ 0 & \text{otherwise} \end{cases}$$

where $k$ is a constant.

**(a)** Show that $k = \dfrac{1}{12}$.         (2 marks)

**(b)** Find the value of $a$ such that $E(aX - 5) = 16.5$.         (4 marks)

## Answer to Question 21

**(a)** Changing the probability function into a table of values:

| $x$ | 1 | 2 | 3 | 4 | 5 |
|---|---|---|---|---|---|
| $P(X = x)$ | $k$ | $2k$ | $2k$ | $3k$ | $4k$ |

Using $\sum p = 1$: $k + 2k + 2k + 3k + 4k = 1$         [M1]

$\qquad\qquad 12k = 1, k = \dfrac{1}{12}$         [A1]

🖉 It is common for grade-C candidates to score 0 for this part, when the answer is given. The most common but incomplete solution looks like this:

$k + 2k + 2k + 3k + 4k = 12k$, so $k = \dfrac{1}{12}$

In this solution there is no mention of the total probability equalling 1. Hence it is an incomplete solution and scores nothing. There is no explanation of how $12k$ suddenly becomes 1/12.

**(b)** $E(X) = \left(1 \times \dfrac{1}{12}\right) + \left(2 \times \dfrac{2}{12}\right) + \left(3 \times \dfrac{2}{12}\right) + \left(4 \times \dfrac{3}{12}\right) + \left(5 \times \dfrac{4}{12}\right) = \dfrac{43}{12}$     [M1, A1]

$E(aX - 5) = aE(X) - 5$     [M1]

$a \times \dfrac{43}{12} - 5 = 16.5$

$a = \dfrac{21.5}{\frac{43}{12}} = 6$     [A1]

🖉 Rounding off the values too early is likely to lose a mark here. If $E(X)$ is rounded to 3.58 then

$\dfrac{21.5}{3.58} = 6.01$

rather than exactly 6. It is much better to use the exact value of $E(X)$.

■ ■ ■

# Question 22

The random variable $X$ has the discrete uniform distribution:

$P(X = x) = \dfrac{1}{7}$, $x = 1, 2, 3, 4, 5, 6, 7$

**(a) Write down the value of $E(X)$.**     (1 mark)
**(b) Calculate the value of Var($X$).**     (2 marks)

# Answer to Question 22

**(a)** $E(X) = 4$     [B1]

🖉 The instruction 'write down' means that no working is needed in order to answer the question. This answer is found by knowing that in a discrete uniform distribution:

$E(X) = \dfrac{1 + n}{2}$, so $\dfrac{1 + 7}{2} = 4$

Grade-C candidates often obtain a correct answer but waste time using the long method:

$E(X) = \left(1 \times \dfrac{1}{7}\right) + \left(2 \times \dfrac{1}{7}\right) + \dots + \left(7 \times \dfrac{1}{7}\right) = 4$

This longer solution would still score the mark.

**(b)** $\text{Var}(X) = \dfrac{n^2 - 1}{12} = \dfrac{7^2 - 1}{12} = \dfrac{48}{12} = 4$ [M1, A1]

✎ Again, the grade-C candidate may well score both marks but perhaps by using another long method, e.g. $\text{Var}(X) = E(X^2) - (E(X))^2$. To save time in the examination, it is worth learning the two rules, as used in the two parts of this question, to find the values of $E(X)$ and $\text{Var}(X)$ for the discrete uniform distribution. The two rules are not given in the formula book.

■ ■ ■

## Question 23

**The random variable Y is normally distributed with a mean $\mu$ and a variance $\sigma^2$.**
**(a) Write down three properties of the distribution of Y.** (3 marks)
**(b) Given that $\mu = 48$ and $\sigma^2 = 100$, find $P(45 < X < 55)$** (5 marks)

## Answer to Question 23

**(a)** It is symmetrical about the mean.
mode = median = mean
It has a bell-shaped curve.
The horizontal axis is an asymptote to the curve.
95% of the data lie within 2 standard deviations of the mean or 68% lie within 1 standard deviation, or almost all, the data lie within 3 standard deviations of the mean
[B1, B1, B1]

✎ Any three of the properties would be accepted. 3 marks = three comments. A sketch of the bell-shaped curve would also be acceptable.

**(b)** $\mu = 48$ and $\sigma^2 = 100$
so $\sigma = 10$

$Z_1 = \dfrac{X_1 - \mu}{\sigma} = \dfrac{45 - 48}{10} = -0.3$ [M1, A1]

$Z_2 = \dfrac{X_2 - \mu}{\sigma} = \dfrac{55 - 48}{10} = 0.7$ [A1]

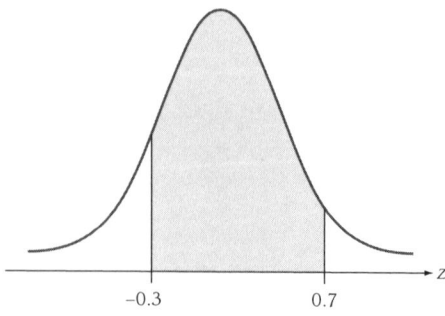

$P(z < 0.7) = 0.7580$
$P(z < -0.3) = 1 - P(z < 0.3) = 1 - 0.6179 = 0.3821$
$P(-0.3 < z < 0.7) = 0.7580 - 0.3821 = 0.3759$ [M1, A1]

📝 Grade-C candidates sometimes use the given variance of 100 in the equation for standardising, instead of the standard deviation of 10. If this happens, then the 2 method marks can still be gained, but all the accuracy marks would be lost.

It is essential that $\sigma$ is used in the standardisation formula, not $\sigma^2$ or $\sqrt{\sigma}$.

■ ■ ■

## Question 24

The light bulbs produced by a particular company have a mean lifetime of 1750 hours with a standard deviation of 40 hours. The lifetimes, *T*, can be assumed to follow a normal distribution.

**(a)** Estimate the percentage of bulbs that could be expected to last longer than 1800 hours. (3 marks)

**(b)** Find the value of t, to 1 decimal place, such that *P(T < t) = 0.15*. (4 marks)

## Answer to Question 24

**(a)** $z = \dfrac{x - \mu}{\sigma} = \dfrac{1800 - 1750}{40} = 1.25$      [M1, A1]

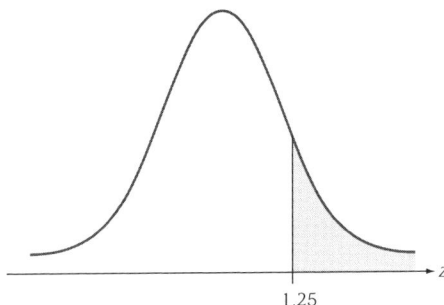

1.25

$P(z > 1.25) = 1 - P(z < 1.25)$
$= 1 - 0.8944$
$= 0.1056$
$= (10.56\%)$      [A1]

📝 This part is usually answered well.

**(b)**

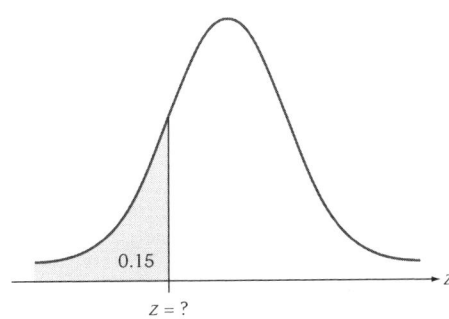

0.15

z = ?

From the 'Percentage points of the normal distribution' table:

$z = -1.0364$                                                                              [B1]

$$-1.0364 = \frac{t - 1750}{40}$$                                                          [M1, A1]

$t = -1.0364 \times 40 + 1750 = 1708.544$

$= 1708.5$ hours (to 1 decimal place)                                                      [A1]

*e* Failing to use the correct table to find the value of z to 4 decimal places is a common mistake made by grade-C candidates, which loses the first mark.

Even more common is for candidates to fail to see that the z value *must* be negative, because the required region is on the left-hand side of the distribution. Drawing the sketch of the curve helps to avoid this second error.

■ ■ ■

## Question 25

**The random variable $X$ is normally distributed with mean $\mu$ and standard deviation $\sigma$. $P(X < 126) = 0.0401$ and $P(X > 160) = 0.0062$. Calculate the value of $\mu$ and the value of $\sigma$.**                          (8 marks)

### Answer to Question 25

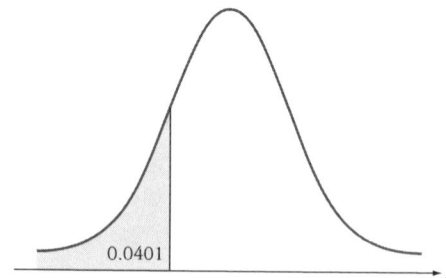

$z_1 = -1.75$ (from normal tables)                                                         [B1]

$$-1.75 = \frac{126 - \mu}{\sigma}$$                                                        [M1]

Equation 1: $\mu - 1.75\sigma = 126$                                                       [A1]

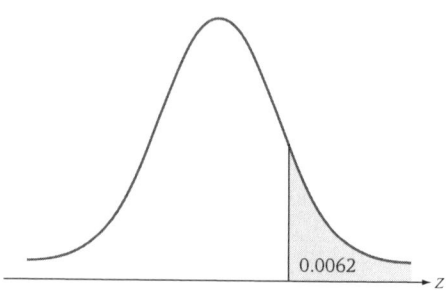

$z_2 = 2.5$        [B1]

$2.5 = \dfrac{160 - \mu}{\sigma}$

Equation 2: $\mu + 2.5\sigma = 160$        [A1]

Solving simultaneously:

    $\mu + 2.5\sigma = 160$

    $\mu - 1.75\sigma = 126$

    $4.25\sigma = 34$        [M1]

    $\sigma = \dfrac{34}{4.25} = 8$        [A1]

    $\mu + 2.5 \times 8 = 160$

    $\mu = 140$        [A1]

    $\mu = 140$ and $\sigma = 8$

✎ Grade-C candidates often have two positive values for the z values, forgetting that z = 0 is the middle of the normal sketch and therefore any values to the left of centre must be *negative* z values.

If the value of +1.75 had been used, then the two answers would be $\mu = 46.667$ and $\sigma = 45.333$.

It should be obvious from the information in the question that $\mu$ must lie between 126 and 160, so the error should be easy to see, with a little thought and checking.